A PLUME BOOK

MY STROKE OF INSIGHT

Kip May

JILL BOLTE TAYLOR, PH.D., is a neuroanatomist affiliated with the Midwest Proton Radiotherapy Institute in Bloomington, Indiana. She is the national spokesperson for the Harvard Brain Tissue Resource Center (Brain Bank) and one of *Time* magazine's 100 Most Influential People in the World, 2008. For more information please visit www.drjilltaylor.com and www.mystrokeofinsight.com.

Praise for *My Stroke of Insight*

"[Dr. Taylor] brings a deep personal understanding to something that she has long studied. . . . Her message, that people can choose to live a more peaceful, spiritual life . . . has resonated widely."   —*The New York Times*

"Provides excellent insights into caring for stroke patients and helping them recover."   —*The Wall Street Journal*

"She has done perhaps more than anyone else to explain, both to the healthy and the stricken, what a stroke is."—*Time*

"Taylor's description of the onset and progression of the stroke . . . is gripping. Taylor's bravery and resilience are formidable and help to dispel the simplistic notion of disability as a disaster. This book is a valuable addition to the narratives of stroke . . . to those in medical humanities programs, as well as to students and professionals in the many disciplines that are involved in the treatment of stroke."
                                    —*The New England Journal of Medicine*

"An instant hit. . . . [Taylor] combines motivational training and bits of spiritualism with the legitimizing language of neuroscience."   —*The Guardian* (London)

"Fascinating . . . Bursts with hope for everyone who is brain injured (not just stroke patients) and gives medical practitioners clear, no-nonsense information about the shortcomings of conventional treatment and attitudes toward the brain injured. . . . But to my mind, what makes *My Stroke of Insight* not just valuable but invaluable—a gift to every spiritual seeker and peace activist—is what I would describe as Taylor's fearless mapping of the physiology of compassion, the physiology of nirvana. This book is about the wonder of being human."   —Tribune Media Services

"Wonderful. . . . [Taylor's] account of her stroke is . . . meticulously detailed."   —*The Huffington Post*

"One of the most significant descriptions of stroke."
                                    —*Long Beach Press-Telegram*

"Jill helps others not only rebuild their brains from trauma, but helps those of us with normal brains better understand how we can . . . maximize our quality of life."

—HealthNewsDigest.com

"Taylor's book explains stroke from the inside. . . . A welcome addition to the literature. A memoir of the transformation. To say [*My Stroke of Insight*] is an intriguing fusion of science and spirituality is an understatement."

—*Sacramento News & Review*

"She has written an important book—fascinating to the layman, but also a significant resource for brain scientists for years to come."

—*Telegraph and Gazette* (Worcester, Massachusetts)

"Moving and informative."   —*Library Journal*

My STROKE *of* INSIGHT

A Brain Scientist's Personal Journey

Jill Bolte Taylor, Ph.D.

A PLUME BOOK

PLUME
Published by the Penguin Group • Penguin Group (USA) Inc., 375 Hudson Street, New York, New York 10014, U.S.A. • Penguin Group (Canada), 90 Eglinton Avenue East, Suite 700, Toronto, Ontario, Canada M4P 2Y3 (a division of Pearson Penguin Canada Inc.) • Penguin Books Ltd., 80 Strand, London WC2R 0RL, England • Penguin Ireland, 25 St. Stephen's Green, Dublin 2, Ireland (a division of Penguin Books Ltd.) • Penguin Group (Australia), 250 Camberwell Road, Camberwell, Victoria 3124, Australia (a division of Pearson Australia Group Pty. Ltd.) • Penguin Books India Pvt. Ltd., 11 Community Centre, Panchsheel Park, New Delhi – 110 017, India • Penguin Group (NZ), 67 Apollo Drive, Rosedale, North Shore 0632, New Zealand (a division of Pearson New Zealand Ltd.) • Penguin Books (South Africa) (Pty.) Ltd., 24 Sturdee Avenue, Rosebank, Johannesburg 2196, South Africa

Penguin Books Ltd., Registered Offices: 80 Strand, London WC2R 0RL, England

Published by Plume, a member of Penguin Group (USA) Inc. Previously published in a Viking edition.

First Plume Printing, June 2009

10 9 8 7 6 5 4 3 2 1

Copyright © Jill Bolte Taylor, 2006
All rights reserved
My Stroke of Insight is a trademark of My Stroke of Insight, Inc.

Photograph of "Stained Glass Brain #3" by Joe LaMantia

 REGISTERED TRADEMARK—MARCA REGISTRADA

ISBN 978-0-670-02074-4 (hc.)
ISBN 978-0-452-29554-4 (pbk.)

Printed in the United States of America

This book is dedicated to G.G.
Thank you, Mama, for helping me heal my mind.
Being your daughter has been my first and greatest blessing.
And to the memory of Nia.
There is no love like puppy love.

ACKNOWLEDGMENTS

My deepest appreciation goes to you, the reader, for sharing this message with your family, friends, and loved ones who value it. I trust that this book will find its way into the hands of those who need it, and you are key to that exchange of information.

I am filled with gratitude for the team of wonderful people who have contributed to the success of *My Stroke of Insight*. My special thanks to my manager Dr. Katherine Domingo, my literary agent/attorney Ellen Stiefler, and my hardworking and generous-of-spirit team at Viking Penguin: Clare Ferraro, Wendy Wolf, Alessandra Lusardi, Carolyn Coleburn, Louise Braverman, Hal Fessenden, Manisha Chakravarthy, and Anna Sternoff. Thank you for helping me bring this message to the world.

CONTENTS

INTRODUCTION
Heart to Heart, Brain to Brain

Every brain has a story and this is mine. Ten years ago, I was at Harvard Medical School performing research and teaching young professionals about the human brain. But on December 10, 1996, I was given a lesson of my own. That morning, I experienced a rare form of stroke in the left hemisphere of my brain. A major hemorrhage, due to an undiagnosed congenital malformation of the blood vessels in my head, erupted unexpectedly. Within four brief hours, through the eyes of a curious brain anatomist (neuroanatomist), I watched my mind completely deteriorate in its ability to process information. By the end of that morning, I could not walk, talk, read, write, or recall any of my life. Curled up into a little fetal ball, I felt my spirit surrender to my death, and it certainly never dawned on me that I would ever be capable of sharing my story with anyone.

My Stroke of Insight: A Brain Scientist's Personal Journey is a chronological documentation of the journey I took into the formless abyss of a silent mind, where the essence of my being became enfolded in a deep inner peace. This book is a

weaving of my academic training with personal experience
and insight. As far as I am aware, this is the first documented
account of a neuroanatomist who has completely recovered
from a severe brain hemorrhage. I am thrilled that these
words will finally go out into the world where they might do
the most good.

More than anything, I am grateful to be alive and celebrative of the time I have here. Initially, I was motivated to endure the agony of recovery thanks to the many beautiful
people who reached out with unconditional love. Over the
years, I have remained obedient to this project because of the
young woman who contacted me out of desperation to understand why her mother, who died from stroke, had not dialed 9-1-1. And because of the elderly gentleman, who was
laden with worry that his wife had suffered enormously while
in a coma before her death. I have remained tethered to my
computer (with my faithful dog Nia on my lap) for the many
caregivers who have called in search of direction and hope. I
have persisted with this work for the seven hundred thousand
people in our society (and their families) who will experience
stroke this year. If just one person reads "Morning of the
Stroke," recognizes the symptoms of stroke, and calls for
help—sooner rather than later—then my efforts over the last
decade will be more than rewarded.

My Stroke of Insight falls into four natural divisions. The
first portion, "Jill's Pre-Stroke Life," introduces you to who I
was before my brain went offline. I describe why I grew up
to be a brain scientist, a little of my academic journey, my advocacy interests, and my personal quest. I was living large. I
was a brain scientist at Harvard, serving on the national board
of NAMI (the National Alliance on Mental Illness), and traveling the country as the *Singin' Scientist*. I follow this very brief

personal synopsis with a little simple science, which is designed to help you understand what was going on in my brain biologically on the morning of the stroke.

If you ever wondered what it might feel like to have a stroke, then the "Morning of the Stroke" chapters are for you. Here, I take you on a very unusual journey into the step-by-step deterioration of my cognitive abilities, as viewed through the eyes of a scientist. As the hemorrhage in my brain grew larger and larger, I relate the cognitive deficits I was experiencing to the underlying biology. As a neuroanatomist, I must say that I learned as much about my brain and how it functions during that stroke, as I had in all my years of academia. By the end of that morning, my consciousness shifted into a perception that I was at *one* with the universe. Since that time, I have come to understand how it is that we are capable of having a "mystical" or "metaphysical" experience—relative to our brain anatomy.

If you know of someone who has had a stroke or some other type of brain trauma, then the recovery chapters may prove to be an invaluable resource. Here, I share the chronological journey of my recovery, including more than fifty tips about things I needed (or didn't need) in order to recover completely. My "Recommendations for Recovery" are listed in the back of the book for your convenience. I hope you will share this information with anyone who may benefit.

Finally, "My Stroke of Insight" defines what this stroke has taught me about my brain. At this point, you will realize that this book is not really about stroke. More accurately, the stroke was the traumatic event through which the insight came. This book is about the beauty and resiliency of our human brain because of its innate ability to constantly adapt to change and recover function. Ultimately, it's about my brain's journey into my right hemisphere's consciousness, where I became enveloped in a deep inner peace. I have resurrected the consciousness of my left hemisphere in order to help others achieve that same inner peace—without having to experience stroke! I hope you enjoy the journey.

My
STROKE
of
INSIGHT

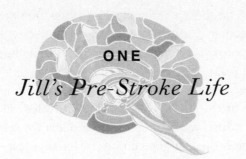

ONE

Jill's Pre-Stroke Life

I am a trained and published neuroanatomist. I grew up in Terre Haute, Indiana. One of my older brothers, who is only eighteen months older than I, was diagnosed with the brain disorder schizophrenia. He was officially diagnosed at the age of thirty-one, but showed obvious signs of psychosis for many years prior to that. During our childhood, he was very different from me in the way he experienced reality and chose to behave. As a result, I became fascinated with the human brain at an early age. I wondered how it could be possible that my brother and I could share the same experience but walk away from the situation with completely different interpretations about what had just happened. This difference in perception, information processing, and output motivated me to become a brain scientist.

My undergraduate academic journey began at Indiana University in Bloomington, Indiana, in the late 1970s. Because of my interactions with my brother, I was hungry to understand what "normal" was at a neurological level. At that time, the subject of neuroscience was such a young field that it was not yet offered on the IU campus as a formal area of specialization. By studying both physiological psychology and human biology, I learned as much as I could about the human brain.

My first real job in the world of medical science turned out to be an enormous blessing in my life. I was hired as a lab technician at the Terre Haute Center for Medical Education (THCME), which is a branch of the Indiana University School of Medicine hosted on the campus of Indiana State University (ISU). My time was evenly divided between the medical human gross anatomy lab and the neuroanatomy research lab. For two years, I was immersed in medical education and, while mentored by Dr. Robert C. Murphy, I fell in love with dissecting the human body.

Bypassing the master's degree, I spent the next six years officially enrolled in the ISU Department of Life Science Ph.D. program. My course load was dominated by the first-year medical school curriculum, and my research specialty was neuroanatomy under the guidance of Dr. William J. Anderson. In 1991, I received my doctorate and felt competent to teach Human Gross Anatomy, Human Neuroanatomy, and Histology at the medical school level.

In 1988, during my time at the THCME and ISU, my brother was officially diagnosed with schizophrenia. Biologically, he is the closest thing to me that exists in the universe. I wanted to understand why I could take my dreams and connect them to reality and make my dreams come true. What was different about my brother's brain such that he could not connect his dreams to a common reality and they instead became delusions? I was eager to pursue research in schizophrenia.

Following commencement from ISU, I was offered a postdoctoral research position at Harvard Medical School in the Department of Neuroscience. I spent two years working with Dr. Roger Tootell on the localization of Area MT, which is located in the part of the brain's visual cortex that tracks movement. I was interested in this project because a high percentage of individuals diagnosed with schizophrenia exhibit abnormal eye behavior when watching moving objects. After helping Roger anatomically identify Area MT's location in the human brain,[1] I followed my heart and transferred into the Harvard Medical School Department of Psychiatry. It was my goal to work in the lab of Dr. Francine M. Benes at McLean Hospital. Dr. Benes is a world-renowned expert in the postmortem investigation of the human brain as it relates to schizophrenia. I believed this would be how I could make my contribution to helping people stricken with the same brain disorder as my brother.

The week before I began my new position at McLean Hospital, my dad, Hal, and I flew to Miami to attend the 1993 annual conference of NAMI, the National Alliance on Mental Illness.[2] Hal, a retired Episcopalian minister with a Ph.D. in counseling psychology, had always been an advocate for social justice. We were both motivated to attend this convention to learn more about NAMI and what we could do to merge our energy with theirs. NAMI is the largest grassroots organization dedicated to improving the lives of persons living with serious mental illness. At that time, NAMI had a membership of approximately 40,000 families with a loved one who is psychiatrically diagnosed. Now NAMI has a membership of approximately 220,000 families. The National NAMI organization advocates at the national level while the State NAMIs advocate at the state level. In addition, there are more than 1,100

1. R. B. H. Tootell and J. B. Taylor, "Anatomical Evidence for MT/V5 and Additional Cortical Visual Areas in Man," in *Cerebral Cortex* (Jan/Feb 1995) 39–55.

2. www.nami.org or 1-800-950-NAMI

local NAMI affiliates scattered around the country providing support, education, and advocacy opportunities for families at the community level.

That trip to Miami changed my life. A group of around fifteen hundred people, comprised of parents, siblings, offspring, and individuals diagnosed with severe mental illness, gathered together for support, education, advocacy, and research-related issues. Until I met other siblings of individuals with mental illness, I had not realized what a profound impact my brother's illness had had on my life. In the course of those few days, I found a family of people who understood the anguish I felt at the loss of my brother to his schizophrenia. They understood my family's struggle to help him access quality treatment. They fought together as an organized voice against the social injustice and stigma related to mental illness. They were armed with educational programs for themselves, as well as for the public, about the biological nature of these disorders. Equally important, they rallied with the brain researchers to help find a cure. I felt that I was in the right place at the right time. I was a sibling, a scientist, and passionate about helping people like my brother. In my soul I felt that I had found not only a cause worthy of my effort, but also an extended family.

The week after the Miami convention, I arrived at McLean Hospital stoked and eager to start my new job in the Laboratory for Structural Neuroscience, the research domain of Dr. Francine Benes. I was all abuzz and thrilled to begin my postmortem investigations into the biological basis of schizophrenia. Francine, whom I affectionately call the "Queen of Schizophrenia," is an amazing research scientist. Just watching how she thinks, how she explores and how she pieces together what she learns from data was a total joy for me. It was a privilege to witness her creativity in experimental design and her persistence, precision, and efficiency in running a research lab. This job was a dream come true. Studying the brains of individuals diagnosed with schizophrenia brought me a feeling of purpose.

On the first day of my new job, however, Francine threw me for a loop when she informed me that the infrequency of brain donations from families of individuals with mental illness had created a long-term shortage of brain tissue for post-mortem investigation. I couldn't believe what I was hearing. I had just spent the better half of a week at National NAMI with hundreds of other families with members diagnosed with severe mental illness. Dr. Lew Judd, a former director of the National Institute of Mental Health, had moderated the research plenary, and several leading scientists had presented their research. NAMI families love sharing and learning about brain research, so I found it mind-boggling that there could be a lack of donated tissue. I decided this was merely a public awareness issue. I believed that once NAMI families knew that there was a research tissue shortage, they would promote brain donation within the organization and resolve the problem.

The next year (1994), I was elected to the National NAMI Board of Directors. It was a thrill for me to be of service to this wonderful organization, a huge honor and responsibility. Of course, the base of my platform was the value of brain donation and the shortage of psychiatrically diagnosed tissue available for scientists to do their work. I called it the "Tissue Issue." At the time, the average age of a NAMI member was sixty-seven years old. I was only thirty-five. I felt proud to be the youngest person ever elected to this board. I had lots of energy and was raring to go.

With my new status within the National NAMI organization, I immediately began keynoting at State NAMI annual conventions all around the country. Before I began this venture, the

Harvard Brain Tissue Resource Center (Brain Bank[3]), which was positioned right next to the Benes Lab, was receiving fewer than three brains a year from psychiatrically diagnosed individuals. This was barely enough tissue for Francine's lab to do its work, much less for the Brain Bank to supply tissue to the other reputable labs that requested it. Within a few months of my traveling around and educating our NAMI families about the "Tissue Issue," the number of brain donations began to increase. Currently, the number of donations from the psychiatrically diagnosed population ranges from twenty-five to thirty-five per year. The scientific community would make good use of a hundred per year.

I realized that early on in my "Tissue Issue" presentations, the subject of brain donation would make some of my audience members squirm uncomfortably. There was this predictable moment when my audience would realize, "Oh my gosh, she wants MY brain!" And I would say to them, "Well yes, yes I do, but don't worry, I'm in no hurry!" To combat their obvious apprehension, I wrote the Brain Bank jingle titled "1-800-BrainBank!"[4] and began traveling with my guitar as the *Singin' Scientist.*[5] As I neared the subject of brain donation and the tension in the room began to rise, I'd pull out my guitar and sing for them. The Brain Bank jingle seems to be just goofy enough to effectively dampen the tension, open hearts, and make it okay for me to communicate my message.

My efforts with NAMI brought deep meaning to my life and my work in the lab flourished. My primary research project in the Benes lab involved working with Francine to create a protocol where we could visualize three neurotransmitter

3. www.brainbank.mclean.org or 1-800-BrainBank
4. See the end of the book for the lyrics to the Brain Bank jingle.
5. www.drjilltaylor.com

systems in the same piece of tissue. Neurotransmitters are the chemicals with which brain cells communicate. This was important work since the newer atypical antipsychotic medications are designed to influence multiple neurotransmitter systems rather than just one. Our ability to visualize three different systems in the same piece of tissue increased our ability to understand the delicate interplay between these systems. It was our goal to better understand the microcircuitry of the brain—which cells in which areas of the brain communicate with which chemicals and in what quantities of those chemicals. The better we understood what the differences were, at a cellular level, between the brains of individuals diagnosed with a severe mental illness and normal control brains, the closer the medical community would be to helping those in need with appropriate medications. In the spring of 1995, this work was featured on the cover of *BioTechniques Journal* and in 1996 it won me the prestigious Mysell Award from the Harvard Medical School Department of Psychiatry. I loved working in the lab and I loved sharing this work with my NAMI family.

And then the unthinkable happened. I was in my midthirties and thriving both professionally and personally. But in one fell swoop, the rosiness of my life and promising future evaporated. I woke up on December 10, 1996, to discover that I had a brain disorder of my own. I was having a stroke. Within four brief hours, I watched my mind completely deteriorate in its ability to process all stimulation coming in through my senses. This rare form of hemorrhage rendered me completely disabled whereby I could not walk, talk, read, write, or recall any aspects of my life.

I realize you are probably eager to begin reading the personal account of the morning of the stroke. However, in order for you to more clearly understand what was going on inside my brain, I have chosen to present some simple science in Chapters Two and Three. Please don't let this section scare you away. I have done my best to keep it user-friendly with lots of simple pictures of the brain so you can understand the anatomy underlying my cognitive, physical, and spiritual experiences. If you absolutely must skip these chapters, then rest assured they will be here for you as a reference. I encourage you to read this section first, however, as I believe it will profoundly simplify your understanding.

TWO

Simple Science

For any two of us to communicate with one another, we must share a certain amount of common reality. As a result, our nervous systems must be virtually identical in their ability to perceive information from the external world, process and integrate that information in our brains, and then have similar systems of output including thought, word, or deed.

The emergence of life was a most remarkable event. With the advent of the single-celled organism, a new era of information processing was born at the molecular level. Through the manipulation of atoms and molecules into DNA and RNA sequences, information could be entered, coded, and stored for future use. Moments in time no longer came and went without a record and, by interweaving a continuum of sequential moments into a common thread, the life of the cell evolved as *a bridge across time*. Before long, cells figured out ways of hanging together and working together, which finally produced you and me.

According to the *American Heritage Dictionary*, to evolve

biologically means "to develop by evolutionary processes from a primitive to a more highly organized form."[6] Earth's molecular brain of DNA is a powerful and successful genetic program—not only because it adapts to constant change, but also because it expects, appreciates, and takes advantage of opportunities to transform itself into something even more magnificent. It is perhaps of interest that our human genetic code is constructed by the exact same four nucleotides (complex molecules) as every other form of life on the planet. At the level of our DNA, we are related to the birds, reptiles, amphibians, other mammals, and even the plant life. From a purely biological perspective, we human beings are our own species-specific mutation of earth's genetic possibility.

As much as we would like to think that human life has attained biological perfection, despite our sophisticated design, we do not represent a finished and/or perfect genetic code. The human brain exists in an ongoing state of change. Even the brains of our ancestors of two thousand or four thousand years ago do not look identical to the brains of man today. The development of language, for example, has altered our brains' anatomical structure and cellular networks.

Most of the different types of cells in our body die and are replaced every few weeks or months. However, neurons, the primary cell of the nervous system, do not multiply (for the most part) after we are born. That means that the majority of the neurons in your brain today are as old as you are. This longevity of the neurons partially accounts for why we feel pretty much the same on the inside at the age of ten as we do

6. Second College Edition (Boston: Houghton Mifflin Company, 1985).

at age thirty or seventy-seven. The cells in our brain are the same, but over time their connections change based upon their/our experience.

The human nervous system is a wonderfully dynamic entity composed of an estimated one trillion cells. To give you some appreciation for how enormous one trillion is, consider this: there are approximately six billion people on the planet and we would have to multiply all six billion people 166 times just to make up the number of cells combining to create a single nervous system!

Of course, our body is much more than a nervous system. In fact, the typical adult human body is composed of approximately fifty trillion cells. That would be 8,333 times all of the six billion people on the planet! What's amazing is that this huge conglomeration of bone cells, muscle cells, connective tissue cells, sensory cells, etc. tend to get along and work together to generate perfect health.

Biological evolution generally occurs from a state of lesser complexity to a state of greater complexity. Nature ensures her own efficiency by not reinventing the wheel with every new species she creates. Generally, once nature identifies a pattern in the genetic code that works toward the survival of the creature, like a blossom for nectar transmission, a heart to pump blood, a sweat gland to help regulate body temperature or an eyeball for vision, she tends to build that feature into future permutations of that specific code. By adding a new level of programming on top of an already well-established set of instructions, each new species contains a strong foundation of time-tested DNA sequences. This is one of the simple ways through which nature transmits the experience and wisdom bestowed by ancient life to her progeny.

Another advantage to this type of build-on-top-of-what-already-works genetic engineering strategy is that very small manipulations of the genetic sequences can result in major evolutionary transformations. In our own genetic profile, believe it or not, scientific evidence indicates that we humans share 99.4 percent of our total DNA sequences with the chimpanzee.[7]

This does not mean, of course, that humans are direct descendants from our tree-swinging friends, but it does emphasize that the genius of our molecular code is supported by eons of nature's greatest evolutionary effort. Our human code was not a random act, at least not in its entirety, but rather is better construed as nature's ever-evolving quest for a body of genetic perfection.

As members of the same human species, you and I share all but 0.01 percent (1/100th of 1 percent) of identical genetic sequences. So biologically, as a species, you and I are virtually identical to one another at the level of our genes (99.99 percent). Looking around at the diversity within our human race, it is obvious that 0.01 percent accounts for a significant difference in how we look, think, and behave.

The portion of our brain that separates us from all other mammals is the outer undulated and convoluted cerebral cortex. Although other mammals do have a cerebral cortex, the human cortex has approximately twice the thickness and is believed to have twice the function. Our cerebral cortex is divided into two major hemispheres, which complement one

7. Derek E. Wildman, et al., Center for Molecular Medicine and Genetics Department of Anatomy and Cell Biology, Wayne State University School of Medicine (Accessed September 10, 2006), <http://www .pnas.org/cgi/content/full/100/12/7181>

another in function. (Note: All of the pictures in this book have the front of the brain directed to the left.)

Whole Human Brain Cerebral Cortex

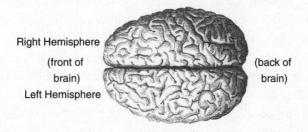

Right Hemisphere
(front of
brain)
Left Hemisphere

(back of
brain)

The two hemispheres communicate with one another through the highway for information transfer, the corpus callosum. Although each hemisphere is unique in the specific types of information it processes, when the two hemispheres are connected to one another, they work together to generate a single seamless perception of the world.

Corpus Callosum
(highway for information transfer)

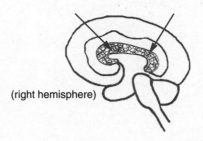

(right hemisphere)

When it comes to the intricate microscopic anatomy of how our cerebral cortices are finely wired, variation is the rule, not the exception. This variation contributes to our individual preferences and personalities. However, the gross (or macroscopic) anatomy of our brains is quite consistent and your brain looks very similar to mine. The bumps (gyri) and grooves (sulci) of the cerebral cortex are specifically organized such that our brains are virtually identical in appearance, structure, and function. For example, each of our cerebral hemispheres contain a superior temporal gyrus, pre- and postcentral gyri, a superior parietal gyrus, along with a lateral occipital gyrus—just to mention a few. Each of these gyri are made up of very specific groups of cells that have very specific connections and functions.

For instance, the cells of the postcentral gyrus enable us to be consciously aware of sensory stimulation, while the cells in the precentral gyrus control our ability to voluntarily move our body parts. The major pathways for information transfer between the various cortical groups of cells (fiber tracts) within each of the two hemispheres are also consistent between us and, as a result, we are generally capable of thinking and feeling in comparable ways.

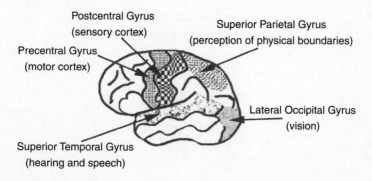

The blood vessels supplying nutrients to our cerebral hemispheres also display a defined pattern. The anterior, middle, and posterior cerebral arteries supply blood to each of the two hemispheres. Damage to any specific branch of one of these major arteries may result in somewhat predictable symptoms of severe impairment or complete elimination of our ability to perform specific cognitive functions. (Of course there are unique differences between damage to the right and left hemispheres.) The following illustration shows the territory of the middle cerebral artery of the left hemisphere, and this includes the location of my stroke. Damage to any of the middle cerebral artery's primary branches would result in relatively predictable symptoms no matter who was having the problem.

Middle Cerebral Artery
Territory and Major Branches

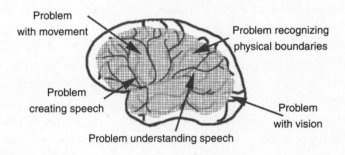

The superficial layers of the cortex, which we see when we look at the external surface of the brain, are filled with neurons that we believe to be uniquely human. These most recently "added on" neurons create circuits that manufacture our ability to think linearly—as in complex language and the ability to think in abstract, symbolic systems like mathematics. The deeper layers of the cerebral cortex make up the cells of the limbic system. These are the cortical cells we share with other mammals.

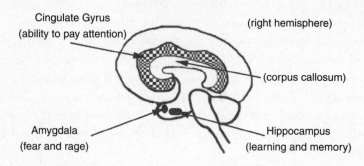

Limbic System
(affect or emotion)

Cingulate Gyrus
(ability to pay attention)

(right hemisphere)

(corpus callosum)

Amygdala
(fear and rage)

Hippocampus
(learning and memory)

The limbic system functions by placing an affect, or emotion, on information streaming in through our senses. Because we share these structures with other creatures, the limbic system cells are often referred to as the "reptilian brain" or the "emotional brain." When we are newborns, these cells become wired together in response to sensory stimulation. It is interesting to note that although our limbic system functions throughout our lifetime, it does not mature. As a result, when our emotional "buttons" are pushed, we retain the ability to react to incoming stimulation as though we were a two-year-old, even when we are adults.

As our higher cortical cells mature and become integrated in complex networks with other neurons, we gain the ability to take "new pictures" of the present moment. When we compare the new information of our thinking mind with the automatic reactivity of our limbic mind, we can reevaluate the current situation and purposely choose a more mature response.

It might be of interest to note that all of today's "brain-based learning" techniques used in elementary through high school capitalize on what neuroscientists understand about the functions of the limbic system. With these learning techniques, we try to transform our classrooms into environments that feel safe and familiar. The objective is to create an environment where the brain's fear/rage response (amygdala) is not triggered. The primary job of the amygdala is to scan all incoming stimulation in this immediate moment and determine the level of safety. One of the jobs of the cingulate gyrus of the limbic system is to focus the brain's attention.

When incoming stimulation is perceived as familiar, the amygdala is calm and the adjacently positioned hippocampus is capable of learning and memorizing new information. However, as soon as the amygdala is triggered by unfamiliar or perhaps threatening stimulation, it raises the brain's level of anxiety and focuses the mind's attention on the immediate situation. Under these circumstances, our attention is shifted away from the hippocampus and focused toward self-preserving behavior about the present moment.

Sensory information streams in through our sensory systems and is immediately processed through our limbic system. By the time a message reaches our cerebral cortex for higher thinking, we have already placed a "feeling" upon how we view that stimulation—is this pain or is this pleasure? Although many of us may think of ourselves as *thinking creatures that feel*, biologically we are *feeling creatures that think*.

Because the term "feeling" is broadly used, I'd like to clarify where different experiences occur in our brain. First, when we experience *feelings* of sadness, joy, anger, frustration, or excitement, these are emotions that are generated by the cells of our limbic system. Second, to *feel* something in your hands refers to the tactile or kinesthetic experience of feeling through the action of palpation. This type of feeling occurs via the sensory system of touch and involves the postcentral gyrus of the cerebral cortex. Finally, when someone contrasts what he or she *feels* intuitively about something (often expressed as a "gut feeling") to what they think about it, this insightful awareness is a higher cognition that is grounded in the right hemisphere of the cerebral cortex. (In Chapter Three we will discuss more thoroughly the different ways in which the right and left cerebral hemispheres operate.)

As information processing machines, our ability to process data about the external world begins at the level of sensory perception. Although most of us are rarely aware of it, our sensory receptors are designed to detect information at the energy level. Because everything around us—the air we breathe, even the materials we use to build with—are composed of spinning and vibrating atomic particles, you and I are literally swimming in a turbulent sea of electromagnetic fields. We are part of it. We are enveloped within it, and through our sensory apparatus we experience *what is*.

Each of our sensory systems is made up of a complex cascade of neurons that process the incoming neural code from the level of the receptor to specific areas within the brain. Each group of cells along the cascade alters or enhances the code, and passes it on to the next set of cells in the system, which further defines and refines the message. By the time the code reaches the outermost portion of our brain, the higher levels of the cerebral cortex, we become conscious of the stimulation. However, if any of the cells along the pathway fail in their ability to function normally, then the final perception is skewed away from normal reality.

Our visual field, the entire view of what we can see when we look out into the world, is divided into billions of tiny spots or pixels. Each pixel is filled with atoms and molecules that are in vibration. The retinal cells in the back of our eyes detect the movement of those atomic particles. Atoms vibrating at different frequencies emit different wavelengths of energy, and this information is eventually coded as different colors by the visual cortex in the occipital region of our brain. A visual image is built by our brain's ability to package groups of pixels together in the form of edges. Different edges with different orientations—vertical, horizontal, and oblique, combine to form complex images. Different groups of cells in our brain add depth, color, and motion to what we see. Dyslexia, whereby some written letters are perceived in reverse from normal, is a great example of a functional abnormality that can occur when the normal cascade of sensory input is altered.

Cortical Organization

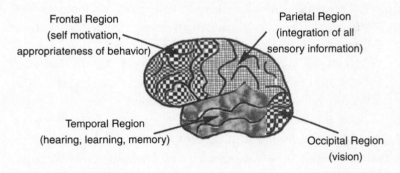

Frontal Region
(self motivation, appropriateness of behavior)

Parietal Region
(integration of all sensory information)

Temporal Region
(hearing, learning, memory)

Occipital Region
(vision)

Similar to vision, our ability to hear sound also depends upon our detection of energy traveling at different wavelengths. Sound is the product of atomic particles in space colliding with one another and emitting patterns of energy. The energy wavelengths, created by the bombarding particles, beat upon the tympanic membrane in our ear. Different wavelengths of sound vibrate our eardrum with unique properties. Similar to our retinal cells, the hair cells of our auditory Organ of Corti translate this energy vibration in our ear into a neural code. This eventually reaches the auditory cortex (in the temporal region of our brain) and we hear sound.

Our most obvious abilities to sense atomic/molecular information occur through our chemical senses of smell and taste. Although these receptors are sensitive to individual electromagnetic particles as they waft past our nose or titillate our taste buds, we are all unique in how much stimulation is required before we can smell or taste something. Each of these sensory systems is also made up of a complex cascade of cells, and damage to any portion of the system may result in an abnormal ability to perceive.

Finally, our skin is our largest sensory organ, and it is stippled with very specific sensory receptors designed to experience pressure, vibration, light touch, pain, or temperature. These receptors are precise in the type of stimulation they perceive such that only cold stimulation can be perceived by cold sensory receptors and only vibration can be detected by vibration receptors. Because of this specificity, our skin is a finely mapped surface of sensory reception.

The innate differences we each experience in terms of how sensitive we are to different types of stimulation contribute greatly to how we perceive the world. If we have problems hearing when people speak, then we will hear only bits and pieces of conversation and make decisions and judgments based upon minimal information. If our eyesight is poor, then we will focus on fewer details and our interaction with the world will be affected. If our sense of smell is deficient, then we may not be able to discriminate between a safe environment and a health hazard, rendering us more vulnerable. At the opposite extreme, if we are oversensitive to stimulation, we may avoid interacting with our environment and miss out on life's simple pleasures.

Pathology and disease of the mammalian nervous system generally involves the brain tissue that distinguishes that specific species from other species. Consequently, in the case of the human system, the outer layers of our cerebral cortex are often vulnerable to disease. Stroke is the number one disabler in our society and the number three killer. Because neurological disease often involves the higher cognition layers of our cerebral cortex, and because stroke occurs four times more frequently in the left cerebral hemisphere, our ability to create or understand language is often compromised. The term "stroke" refers to a problem with the blood vessels carrying oxygen to the cells of the brain, and there are basically two types: ischemic (ih-skee-mik) and hemorrhagic (hem-o-radg-ik).

According to the American Stroke Association, the ischemic stroke accounts for approximately 83 percent of all strokes. Arteries carry blood into the brain and their shape tapers smaller and smaller as they travel farther away from the heart. These arteries carry life-supporting oxygen necessary for cells, including neurons, to survive. With ischemic stroke, a blood clot travels into the artery until the tapered diameter of the artery becomes too small for the clot to pass any farther. The blood clot blocks the flow of oxygen-rich blood to the cells beyond the point of obstruction. Consequently, brain cells become traumatized and often die. Since neurons generally do not regenerate, the dead neurons are not replaced. The function of the deceased cells may be lost permanently, unless other neurons adapt over time to carry out their function. Because every brain is unique in its neurological wiring, every brain is unique in its ability to recover from trauma.

Ischemic Blood Clot

(artery is blocked and oxygen cannot get to cells)

The hemorrhagic stroke occurs when blood escapes from the arteries and floods into the brain. Seventeen percent of all strokes are hemorrhagic. Blood is toxic to neurons when it comes in direct contact with them, so any leak or vascular blowout can have devastating effects on the brain. One form of stroke, the aneurysm (an-yu-rism), forms when there is a weakening in the wall of a blood vessel that consequently balloons out. The weakened area fills with blood and can readily rupture, spewing large volumes of blood into the skull. Any type of hemorrhage is often life threatening.

Aneurysm
(thin wall of blood vessel ballooning out)

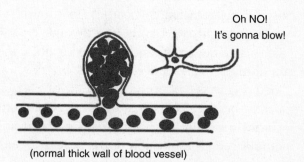

(normal thick wall of blood vessel)

An arteriovenous malformation (AVM) is a rare form of hemorrhagic stroke. It is a congenital disorder whereby an individual is born with an abnormal arterial configuration. Normally, the heart pumps blood through arteries with high pressure while blood is retrieved through veins, which are low pressure. A capillary bed acts as a buffering system or neutral zone between the high-pressure arteries and the low-pressure veins.

Normal Blood Flow

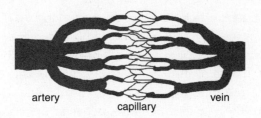

artery   vein
                 capillary

In the case of the AVM, an artery is directly connected to a vein with no buffering capillary bed in between. Over time, the vein can no longer handle the high pressure from the artery and the connection between the artery and vein is broken—spilling blood into the brain. Although the AVM accounts for only 2 percent of all hemorrhagic strokes,[8] it is the most common form of stroke that strikes people during their prime years of life (ages twenty-five to forty-five). I was thirty-seven when my AVM blew.

8. National Institute of Neurological Disorders and Stroke (Accessed September 10, 2006), <http://www.ninds.nih.gov>

Arteriovenous Malformation
(AVM)

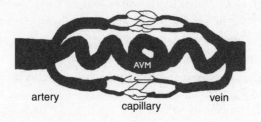

artery   vein
capillary

Regardless of the mechanical nature of the vascular problem, be it a blood clot or a hemorrhage, no two strokes are
identical in their symptoms because no two brains are absolutely identical in their structure, connections, or ability to recover. At the same time, it is impossible to talk about the
symptoms resulting from stroke without having a conversation about the innate differences between the right and left
cerebral hemispheres. Although the anatomical structure of
the two hemispheres is relatively symmetrical, they are quite
diverse in not only how they process information, but also in
the types of information they process.

The better we understand the functional organization of
the two cerebral hemispheres, the easier it is to predict what
deficits might occur when specific areas are damaged. Perhaps more important, we might gain some insight into what
we can do to help stroke survivors recover lost function.

WARNING SIGNS OF STROKE

S = SPEECH, or any problems with language

T = TINGLING, or any numbness in the body

R = REMEMBER, or any problems with memory

O = OFF BALANCE, problems with coordination

K = KILLER HEADACHE

E = EYES, or any problems with vision

STROKE is a medical emergency.

Call 9-1-1.

THREE

Hemispheric Asymmetries

Scientists have been studying the functional asymmetries of the human cerebral cortices for more than two hundred years. To my knowledge, the first person on record to suggest that each hemisphere actually had its own mind was Meinard Simon Du Pui. In 1780, Du Pui claimed that mankind was *Homo Duplex*—meaning that he had a double brain with a double mind.[9] Nearly a century later, in the late 1800s, Arthur Ladbroke Wigan witnessed the autopsy of a man who could walk, talk, read, write, and function like a normal man. Upon examination of his brain, however, Wigan discovered that this man had only one cerebral hemisphere. Wigan concluded that since this man, who had only "half" a brain, had a whole mind and could function like a whole man, then those of us who have two hemispheres must have two minds. Wigan enthusiastically championed this "Duality of the Mind" theory.[10]

9. G. J. C. Lokhorst's *Hemisphere Differences Before 1800* (Accessed September 10, 2006), <http://homepages.ipact.nl/~lokhorst/bbs1985.html>

10. Ibid.

Over the centuries, various conclusions have been drawn about the differences and similarities in how the two hemispheres process information and learn new material. This subject gained tremendous popularity in the United States in the 1970s, following a series of split-brain experiments where Dr. Roger W. Sperry surgically cut the fibers of the corpus callosum of people experiencing severe epileptic seizures. In his 1981 Nobel lecture, Sperry commented:

> Under the conditions of commissurotomy where background factors are equalized and where close left-right comparisons become possible within the same subject working the same problem, even slight lateral differences become significant. The same individual can be observed to employ consistently one or the other of two distinct forms of mental approach and strategy, much like two different people, depending on whether the left or right hemisphere is in use.[11]

Since those early studies of split-brain patients, neuroscientists have learned that the two hemispheres perform differently when they are connected to one another than when they are surgically separated.[12] When normally connected, the two hemispheres complement and enhance one another's abilities. When surgically separated, the two hemispheres function as two independent brains with unique

11. Roger W. Sperry's December 8, 1981, lecture (Accessed on September 10, 2006), <http://nobelprize.org/nobel_prizes/medicine/laureates/1981/sperry-lecture.html>

12. Sperry, M. S. Gazzaniga, and J. E. Bogen, "Interhemispheric Relationships: The Neurocortical Commissures; Syndromes of Hemisphere Disconnection" in *Handbook of Clinical Neurology*, P. J. Vinken and G. W. Bruyn, eds. (Amsterdam: North-Holland Publishing, 1969), 177–84.

personalities, often described as the Dr. Jekyll and Mr. Hyde phenomenon.

Using noninvasive modern techniques including functional imaging (fMRI), scientists are now capable of visualizing which specific neurons are engaged in performing a designated function in real time. Because our two hemispheres are so neuronally integrated via the corpus callosum, virtually every cognitive behavior we exhibit involves activity in both hemispheres—they simply do it differently. As a result, the world of science supports the idea that the relationship between the two cerebral hemispheres is more appropriately viewed as two complementary halves of a whole rather than as two individual entities or identities.

It makes sense that having two cerebral hemispheres that process information in uniquely different ways would increase our brain's capacity to experience the world around us and increase our chances of survival as a species. Because our two hemispheres are so adept at weaving together a single seamless perception of the world, it is virtually impossible for us to consciously distinguish between what is going on in our left hemisphere versus our right hemisphere.

To begin, it is important to understand that hemispheric dominance is not to be confused with hand dominance. Dominance in the brain is determined by which hemisphere houses the ability to create and understand verbal language. Although the statistics vary depending upon whom you ask, virtually everyone who is right-handed (more than 85 percent of the U.S. population) is left hemisphere dominant. At the same time, more than 60 percent of left-handed people are also classified as left hemisphere dominant. Let's take a closer look at the asymmetries of the two hemispheres.

Our right hemisphere (which controls the left half of our

body) functions like a parallel processor. Independent streams of information simultaneously burst into our brain via each of our sensory systems. Moment by moment, our right mind creates a master collage of what this moment in time looks like, sounds like, tastes like, smells like, and feels like. Moments don't come and go in a rush, but rather are rich with sensations, thoughts, emotions, and often, physiological responses. Information processed in this way allows us to take an immediate inventory about the space around us and our relationship to that space.

Thanks to the skills of our right mind, we are capable of remembering isolated moments with uncanny clarity and accuracy. Most of us can remember where we were and how we felt when we first heard about the assassination of President Kennedy or saw the collapse of the World Trade Center. Do you remember the moment you spoke the words "I do," or first saw your newborn smile? Our right hemisphere is designed to remember things as they relate to one another. Borders between specific entities are softened, and complex mental collages can be recalled in their entirety as combinations of images, kinesthetics, and physiology.

To the right mind, no time exists other than the present moment, and each moment is vibrant with sensation. Life or death occurs in the present moment. The experience of joy happens in the present moment. Our perception and experience of connection with something that is greater than ourselves occurs in the present moment. To our right mind, the moment of *now* is timeless and abundant.

In the absence of all the rules and regulations that have already been defined as the correct way of doing something, our right mind is free to think intuitively outside the box, and it creatively explores the possibilities that each new moment brings. By its design, our right mind is spontaneous, carefree, and imaginative. It allows our artistic juices to flow free without inhibition or judgment.

The present moment is a time when everything and everyone are connected together as *one*. As a result, our right mind perceives each of us as equal members of the human family. It identifies our similarities and recognizes our relationship with this marvelous planet, which sustains our life. It perceives the big picture, how everything is related, and how we all join together to make up the whole. Our ability to be empathic, to walk in the shoes of another and feel their feelings, is a product of our right frontal cortex.

In contrast, our left hemisphere is completely different in the way it processes information. It takes each of those rich and complex moments created by the right hemisphere and strings them together in timely succession. It then sequentially compares the details making up this moment with the details making up the last moment. By organizing details in a linear and methodical configuration, our left brain manifests the concept of time whereby our moments are divided into the past, present, and future. Within the structure of this predictable temporal cadence, we can appreciate that this must occur before that can happen. I look at my shoes and socks and it is my left hemisphere that comprehends that I must put my socks on before my shoes. It can look at all the details of a puzzle and use the clues of color, shape, and size to recognize patterns for arrangement. It builds an understanding of everything using deductive reasoning such that if A is greater than B, and B is greater than C, then A must be greater than C.

Just opposite to how our right hemisphere thinks in pictures and perceives the big picture of the present moment, our left mind thrives on details, details, and more details about those details. Our left hemisphere language centers use words to describe, define, categorize, and communicate about everything. They break the big picture perception of the present moment into manageable and comparable bits of data that they can talk about. Our left hemisphere looks at a flower and names the different parts making up the whole—the petal, stem, stamen, and pollen. It dissects the image of a rainbow into the language of red, orange, yellow, green, blue, indigo, and violet. It describes our body as arms, legs, a torso, and every anatomical, physiological, and biochemical detail one can imagine. It thrives on weaving facts and details into a story. It excels in academics, and by doing so, it manifests a sense of authority over the details it masters.

Via our left hemisphere language centers, our mind speaks to us constantly, a phenomenon I refer to as "brain chatter." It is that voice reminding you to pick up bananas on your way home and that calculating intelligence that knows when you have to do your laundry. There is vast individual variation in the speed at which our minds function. For some, our dialogue of brain chatter runs so fast that we can barely keep up with what we are thinking. Others of us think in language so slowly that it takes a long time for us to comprehend. Still others of us have a problem retaining our focus and concentration long enough to act on our thoughts. These variations in normal processing stem back to our brain cells and how each brain is intrinsically wired.

One of the jobs of our left hemisphere language centers is to define our *self* by saying "I am." Through the use of brain chatter, your brain repeats over and over again the details of your life so you can remember them. It is the home of your ego center, which provides you with an internal awareness of what your name is, what your credentials are, and where you live. Without these cells performing their job, you would forget who you are and lose track of your life and your identity.

Along with thinking in language, our left hemisphere thinks in patterned responses to incoming stimulation. It establishes neurological circuits that run relatively automatically to sensory information. These circuits allow us to process large volumes of information without having to spend much time focusing on the individual bits of data. From a neurological standpoint, every time a circuit of neurons is stimulated, it takes less external stimulation for that particular circuit to run. As a result of this type of reverberating circuitry, our left hemisphere creates what I call "loops of thought patterns" that it uses to rapidly interpret large volumes of incoming stimulation with minimal attention and calculation.

Because our left brain is filled with these ingrained programs of pattern recognition, it is superb at predicting what we will think, how we will act, or what we will feel in the future—based upon our past experience. I, personally, love the color red and am inclined to collect a bunch of red things—I drive a red car and wear red clothes. I like red because there's a circuit in my brain that gets very excited and runs relatively automatically when anything red comes my way. From a purely neurological perspective, I like red because the cells in my left brain tell me I like red.

Among other things, our left hemisphere categorizes information into hierarchies, including things that attract us (our likes) or repel us (our dislikes). It places the judgment of good on those things we like and bad on those things we dislike. Through the action of critical judgment and analysis, our left brain constantly compares us with everyone else. It keeps us abreast of where we stand on the financial scale, academic scale, honesty scale, generosity-of-spirit scale, and every other scale you can imagine. Our ego mind revels in our individuality, honors our uniqueness, and strives for independence.

*   *   *

Although each of our cerebral hemispheres process information in uniquely different ways, the two work intimately with one another when it comes to just about every action we undertake. With language, for example, our left hemisphere understands the details making up the structure and semantics of the sentence—and the meaning of the words. It is our left mind that understands what letters are and how they fit together to create a sound (word) that has a concept (meaning) attached to it. It then strings words together in a linear fashion to create sentences and paragraphs capable of conveying very complex messages.

Our right hemisphere complements the action of our left hemisphere language centers by interpreting nonverbal communication. Our right mind evaluates the more subtle cues of language, including tone of voice, facial expression, and body language. Our right hemisphere looks at the big picture of communication and assesses the congruity of the overall expression. Any inconsistencies between how someone holds their body, versus their facial expression, versus their tone of voice, versus the message they are communicating, might indicate either a neurological abnormality in how someone expresses himself or it may prove to be a telltale sign that the person is not telling the truth.

People who have damage in their left hemisphere often cannot create or understand speech because the cells in their language centers have been injured. However, they are often genius at being able to determine if someone is telling the truth, thanks to the cells in their right hemisphere. On the other hand, if someone has damage to their right hemisphere, they may not appropriately assess the emotional content of a message. For example, if I am playing blackjack at a party and I say, "Hit me!" a person with a damaged right hemisphere may think I am asking him to physically strike me rather than understand that I am simply asking for another card. Without the right hemisphere's ability to evaluate communication in the context of the bigger picture, the left hemisphere tends to interpret everything literally.

Music is another great example of how our two hemispheres complement one another in function. When we methodically and meticulously drill our scales over and over again, when we learn to read the language of staff notation, and when we memorize which fingering on an instrument will create which named note, we are tapping primarily into the skills of our left brain. Our right brain kicks into high gear when we are doing things in the present moment—like performing, improvising, or playing by ear.

Did you ever stop to consider how it is that your brain knows how to define the dimensions of your body in space? Amazingly, there are cells in our left hemisphere's orientation association area that define the boundaries of our body—where we begin and where we end relative to the space around us. At the same time, there are cells in our right hemisphere's orientation association area that orient our body in space. As a result, our left hemisphere teaches us where our body begins and ends, and our right hemisphere helps us place it where we want it to go.[13]

I enthusiastically encourage you to explore myriad current literature about teaching and the brain, learning and the brain, and the asymmetries of our two cortical hemispheres. I believe that the more we understand about how our hemispheres work together to create our perception of reality, then the more successful we will be in understanding the natural

13. Andrew Newberg, Eugene D'Aquili, and Vince Rause, *Why God Won't Go Away* (NY: Ballantine, 2001), 28.

gifts of our own brains, as well as more effectively help people recover from neurological trauma.

The type of stroke I experienced was a severe hemorrhage in the left hemisphere of my brain due to an undiagnosed AVM. On the morning of the stroke, this massive hemorrhage rendered me so completely disabled that I describe myself as an infant in a woman's body. Two and a half weeks after the stroke, I underwent major surgery to remove a golf ball–sized blood clot that was obstructing my brain's ability to transmit information.

Following surgery, it took eight years for me to completely recover all physical and mental functions. I believe I have recovered completely because I had an advantage. As a trained neuroanatomist, I believed in the plasticity of my brain—its ability to repair, replace, and retrain its neural circuitry. In addition, thanks to my academics, I had a "roadmap" to understanding how my brain cells needed to be treated in order for them to recover.

The story that follows is my stroke of insight into the beauty and resiliency of the human brain. It's a personal account, as seen through the eyes of a neuroscientist, about what it felt like to experience the deterioration of my left brain and then recover it. It is my hope that this book will offer insight into how the brain works in both wellness and in illness. Although this book is written for the general public, I hope you will share it with people you want to help recover from brain trauma and their caregivers.

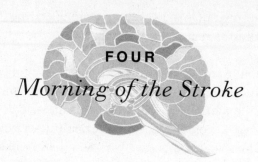

FOUR

Morning of the Stroke

It was 7:00 A.M. on December 10, 1996. I awoke to the familiar tick-tick-tick of my compact disc player as it began winding up to play. Sleepily, I hit the snooze button just in time to catch the next mental wave back into dreamland. Here, in this magic land I call "Thetaville"—a surreal place of altered consciousness somewhere between dreams and stark reality—my spirit beamed beautiful, fluid, and free from the confines of normal reality.

Six minutes later, as the tick-tick-tick of the CD alerted my memory that I was a land mammal, I sluggishly awoke to a sharp pain piercing my brain directly behind my left eye. Squinting into the early morning light, I clicked off the impending alarm with my right hand and instinctively pressed the palm of my left hand firmly against the side of my face. Rarely ill, I thought how queer it was for me to awaken to such a striking pain. As my left eye pulsed with a slow and deliberate rhythm, I felt bewildered and irritated. The throbbing pain behind my eye was sharp, like the caustic sensation that sometimes accompanies biting into ice cream.

As I rolled out of my warm waterbed, I stumbled into the world with the ambivalence of a wounded soldier. I closed the bedroom window blind to block the incoming stream of light from stinging my eyes. I decided that exercise might get my blood flowing and perhaps help dissipate the pain. Within moments, I hopped on to my "cardio-glider" (a full body exercise machine) and began jamming away to Shania Twain singing the lyrics, "Whose bed have your boots been under?" Immediately, I felt a powerful and unusual sense of dissociation roll over me. I felt so peculiar that I questioned my well-being. Even though my thoughts seemed lucid, my body felt irregular. As I watched my hands and arms rocking forward and back, forward and back, in opposing synchrony with my torso, I felt strangely detached from my normal cognitive functions. It was as if the integrity of my mind/body connection had somehow become compromised.

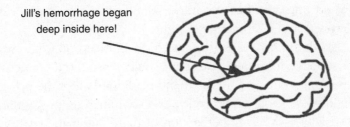

Jill's hemorrhage began
deep inside here!

Feeling detached from normal reality, I seemed to be witnessing my activity as opposed to feeling like the active participant performing the action. I felt as though I was observing myself in motion, as in the playback of a memory. My fingers, as they grasped on to the handrail, looked like primitive claws. For a few seconds I rocked and watched, with riveting wonder, as my body oscillated rhythmically and mechanically. My torso moved up and down in perfect cadence with the music and my head continued to ache.

I felt bizarre, as if my conscious mind was suspended somewhere between my normal reality and some esoteric space. Although this experience was somewhat reminiscent of my morning time in Thetaville, I was sure that this time I was awake. Yet, I felt as if I was trapped inside the perception of a meditation that I could neither stop nor escape. Dazed, I felt the frequency of shooting pangs escalate inside my brain, and I realized that this exercise regime was probably not a good idea.

Feeling a little nervous about my physical condition, I climbed off the machine and bumbled through my living room on the way to the bath. As I walked, I noticed that my movements were no longer fluid. Instead they felt deliberate and almost jerky. In the absence of my normal muscular co-ordination, there was no grace to my pace and my balance was so impaired that my mind seemed completely preoccupied with just keeping me upright.

As I lifted my leg to step into the tub, I held on to the wall for support. It seemed odd that I could sense the inner activities of my brain as it adjusted and readjusted all of the opposing muscle groups in my lower extremities to prevent me from falling over. My perception of these automatic body responses was no longer an exercise in intellectual conceptualization. Instead, I was momentarily privy to a precise and experiential understanding of how hard the fifty trillion cells in my brain and body were working in perfect unison to maintain the flexibility and integrity of my physical form. Through the eyes of an avid enthusiast of the magnificence of the human design, I witnessed with awe the autonomic functioning of my nervous system as it calculated and recalculated every joint angle.

Ignorant to the degree of danger my body was in, I balanced my weight against the shower wall. As I leaned forward to turn on the faucet, I was startled by an abrupt and exaggerated clamor as water surged into the tub. This unexpected amplification of sound was both enlightening and disturbing. It brought me to the realization that, in addition to having problems with coordination and equilibrium, my ability to process incoming sound (auditory information) was erratic.

I understood neuroanatomically that coordination, equilibrium, audition, and the action of inspirational breathing were processed through the pons of my brainstem. For the first time, I considered the possibility that I was perhaps having a major neurological malfunction that was life threatening.

Fibers Passing Through the Pons of the Brainstem

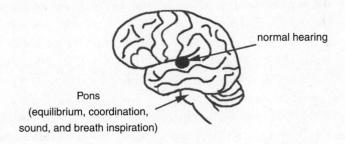

As my cognitive mind searched for an explanation about what was happening anatomically inside my brain, I reeled backward in response to the augmented roar of the water as the unexpected noise pierced my delicate and aching brain.

In that instant, I suddenly felt vulnerable, and I noticed that the constant brain chatter that routinely familiarized me with my surroundings was no longer a predictable and constant flow of conversation. Instead, my verbal thoughts were now inconsistent, fragmented, and interrupted by an intermittent silence.

Language Centers

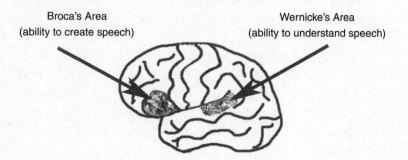

Broca's Area
(ability to create speech)

Wernicke's Area
(ability to understand speech)

When I realized that the sensations outside of me, including the remote sounds of a bustling city beyond my apartment window, had faded away, I could tell that the broad range of my natural observation had become constricted. As my brain chatter began to disintegrate, I felt an odd sense of isolation. My blood pressure must have been dropping as a result of the bleeding in my brain because I felt as if all of my systems, including my mind's ability to instigate movement, were moving into a slow mode of operation. Yet, even though my thoughts were no longer a constant stream of chatter about the external world and my relationship to it, I was conscious and constantly present within my mind.

Confused, I searched the memory banks of both my body and brain, questioning and analyzing anything I could remember having experienced in the past that was remotely similar to this situation. *What is going on?* I wondered. *Have I ever experienced anything like this before? Have I ever felt like this before? This feels like a migraine. What is happening in my brain?*

The harder I tried to concentrate, the more fleeting my ideas seemed to be. Instead of finding answers and information, I met a growing sense of peace. In place of that constant chatter that had attached me to the details of my life, I felt enfolded by a blanket of tranquil euphoria. How fortunate I was that the portion of my brain that registered fear, my amygdala, had not reacted with alarm to these unusual circumstances and shifted me into a state of panic. As the language centers in my left hemisphere grew increasingly silent and I became detached from the memories of my life, I was comforted by an expanding sense of grace. In this void of higher cognition and details pertaining to my normal life, my consciousness soared into an all-knowingness, a "being at *one*" with the universe, if you will. In a compelling sort of way, it felt like the good road home and I liked it.

By this point I had lost touch with much of the physical three-dimensional reality that surrounded me. My body was propped up against the shower wall and I found it odd that I was aware that I could no longer clearly discern the physical boundaries of where I began and where I ended. I sensed the composition of my being as that of a fluid rather than that of a solid. I no longer perceived myself as a whole object separate from everything. Instead, I now blended in with the space and flow around me. Beholding a growing sense of detachment between my cognitive mind and my ability to control and finely manipulate my fingers, the mass of my body felt heavy and my energy waned.

Orientation Association Area
(physical boundaries, space, and time)

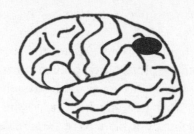

When the shower droplets beat into my chest like little bullets, I was harshly startled back into this reality. As I held my hands up in front of my face and wiggled my fingers, I was simultaneously perplexed and intrigued. *Wow, what a strange and amazing thing I am. What a bizarre living being I am. Life! I am life! I am a sea of water bound inside this membranous pouch. Here, in this form, I am a conscious mind and this body is the vehicle through which I am ALIVE! I am trillions of cells sharing a common mind. I am here, now, thriving as life. Wow! What an unfathomable concept! I am cellular life, no—I am molecular life with manual dexterity and a cognitive mind!*

In this altered state of being, my mind was no longer pre-occupied with the billions of details that my brain routinely used to define and conduct my life in the external world. Those little voices, that brain chatter that customarily kept me abreast of myself in relation to the world outside of me, were delightfully silent. And in their absence, my memories of the past and my dreams of the future evaporated. I was alone. In the moment, I was alone with nothing but the rhythmic pulse of my beating heart.

I must admit that the growing void in my traumatized brain was entirely seductive. I welcomed the reprieve that the silence brought from the constant chatter that related me to what I now perceived as the insignificant affairs of society. I eagerly turned my focus inward to the steadfast drumming of the trillions of brilliant cells that worked diligently and synchronously to maintain my body's steady state of homeostasis. As the blood poured in over my brain, my consciousness slowed to a soothing and satisfying awareness that embraced the vast and wondrous world within. I was both fascinated and humbled by how hard my little cells worked, moment by moment, just to maintain the integrity of my existence in this physical form.

For the first time, I felt truly at one with my body as a complex construction of living, thriving organisms. I was proud to see that I was this swarming conglomeration of cellular life that had stemmed from the intelligence of a single molecular genius! I welcomed the opportunity to pass beyond my normal perceptions, away from the persevering pain that relentlessly pulsed in my head. As my consciousness slipped into a state of peaceful grace, I felt ethereal. Although the pulse of pain in my brain was inescapable, it was not debilitating.

Standing there with the water pounding onto my breasts, a tingling sensation surged through my chest and forcefully radiated upward into my throat. Startled, I became instantly aware that I was in grave danger. Shocked back into this external reality, I immediately reassessed the abnormalities of my physical systems. Determined to understand what was going on, I actively scanned my reservoir of education in demand of a self-diagnosis. *What is going on with my body? What is wrong with my brain?*

Although the sporadically discontinuous flow of normal cognition was virtually incapacitating, somehow I managed to keep my body on task. Stepping out of the shower, my brain felt inebriated. My body was unsteady, felt heavy, and exerted itself in very slow motion. *What is it I'm trying to do? Dress, dress for work. I'm dressing for work.* I labored mechanically to choose my clothes and by 8:15 A.M., I was ready for my commute. Pacing my apartment, I thought, *Okay, I'm going to work. I'm going to work. Do I know how to get to work? Can I drive?* As I visualized the road to McLean Hospital, I was literally thrown off balance when my right arm dropped completely paralyzed against my side. In that moment I knew. *Oh my gosh, I'm having a stroke! I'm having a stroke!* And in the next instant, the thought flashed through my mind, *Wow, this is so cool!*

I felt as though I was suspended in a peculiar euphoric stupor, and I was strangely elated when I understood that this unexpected pilgrimage into the intricate functions of my brain actually had a physiological basis and explanation. I kept thinking, *Wow, how many scientists have the opportunity to study their own brain function and mental deterioration from the inside out?* My entire life had been dedicated to my own understanding of how the human brain creates our perception of reality. And now I was experiencing this most remarkable stroke of insight!

When my right arm became paralyzed, I felt the life force inside the limb explode. When it dropped dead against my body, it clubbed my torso. It was the strangest sensation. I felt as if my arm had been guillotined off!

Movement and Sensory Perception

Motor Cortex
(ability to move)

Sensory Cortex
(ability to sense the world)

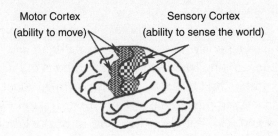

I understood neuroanatomically that my motor cortex had been affected and I was fortunate that within a few minutes, the deadness of my right arm subtly abated. As the limb began to reclaim its life, it throbbed with a formidable tingling pain. I felt weak and wounded. My arm felt completely depleted of its intrinsic strength, yet I could wield it like a stub. I wondered if it would ever be normal again. Catching sight of my warm and cradling waterbed, I seemed to be beckoned by it on this cold winter morning in New England. *Oh, I am so tired. I feel so tired. I just want to rest. I just want to lie down and relax for a little while.* But resounding like thunder from deep within my being, a commanding voice spoke clearly to me: *If you lie down now you will never get up!*

Startled by this ominous illumination, I fathomed the gravity of my immediate situation. Although I was compelled by a sense of urgency to orchestrate my rescue, another part of me delighted in the euphoria of my irrationality. I stepped across the threshold of my bedroom, and as I gazed into the eyes of my reflected image, I paused for a moment, in search of some guidance or profound insight. In the wisdom of my dementia, I understood that my body was, by the magnificence of its biological design, a precious and fragile gift. It was clear to me that this body functioned like a portal through which the energy of who I am can be beamed into a three-dimensional external space.

This cellular mass of my body had provided me with a marvelous temporary home. This amazing brain had been capable of integrating literally billions of trillions of bits of data, in every instant, to create for me a three-dimensional perception of this environment that actually appeared to be not only seamless and real, but also safe. Here in this delusion, I was mesmerized by the efficiency of this biological matrix as it created my form, and I was awed by the simplicity of its design. I saw myself as a complex composite of dynamic systems, a collection of interlacing cells capable of integrating a medley of sensory modalities streaming in from the external world. And when the systems functioned properly, they naturally manifested a consciousness capable of perceiving a normal reality. I wondered how I could have spent so many years in this body, in this form of life, and never really understood that I was just visiting here.

Even in this condition, the egotistical mind of my left hemisphere arrogantly retained the belief that although I was experiencing a dramatic mental incapacity, my life was invincible. Optimistically, I believed that I would recover completely from this morning's events. Feeling a little irritated by this impromptu disruption of my work schedule, I bantered, *Okay, well, I'm having a stroke. Yep, I'm having a stroke . . . but I'm a very busy woman! All right, since I can't stop this stroke from happening, then, okay, I'll do this for a week! I'll learn what I need to know about how my brain creates my perception of reality and then I'll meet my schedule, next week. Now, what am I doing? Getting help. I must stay focused and get help.*

Territory of Jill's Hemorrhage
(shaded oval area)

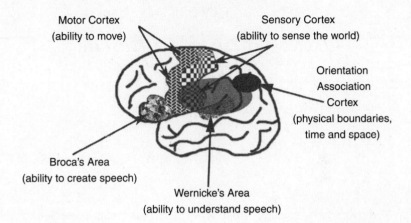

Motor Cortex
(ability to move)

Sensory Cortex
(ability to sense the world)

Orientation
Association
Cortex
(physical boundaries,
time and space)

Broca's Area
(ability to create speech)

Wernicke's Area
(ability to understand speech)

To my counterpart in the looking glass I pleaded, *Remember, please remember everything you are experiencing! Let this be my stroke of insight into the disintegration of my own cognitive mind.*

FIVE

Orchestrating My Rescue

I didn't know exactly what type of stroke I was experiencing, but the congenital arteriovenous malformation (AVM) that burst in my head was spewing a large volume of blood over the left hemisphere of my brain. As blood swept over the higher thinking centers of my left cerebral cortex, I began losing my skills of higher cognition—one precious ability at a time. It was fortunate that I could remember that the best prognosis for someone having a stroke was to get him or her to the hospital as quickly as possible. But getting help was challenging because I found it almost impossible to concentrate or keep my mind on task. I caught myself chasing random thoughts as they danced in and out of my brain, and sadly, I was fully aware that I was inept at holding a plan in my mind long enough to execute it.

The two cerebral hemispheres of my brain had worked meticulously well together for my entire life, as they enabled me to function in the world. But now, because of the normal differences and asymmetry of function between my right and left hemispheres, I felt disjoined from the linguistic and calculating skills of my left brain. Where were my numbers? Where was my language . . . what had become of the brain chatter, which was now replaced by a pervasive and enticing inner peace?

CT Image of Jill's Brain on the Morning of the Stroke

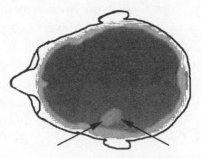

(Hemorrhage in Jill's left hemisphere)

Without the linearity associated with the constant brain directives of my left brain, I struggled to maintain a cognitive connection to my external reality. Instead of a continuous flow of experience that could be divided into past, present, and future, every moment seemed to exist in perfect isolation. In this emptiness of verbal cues, I felt devoid of my worldly wisdom and I was desperate to maintain a cognitive link between my moments. Repetitively, I obsessed the only message my brain could sustain: *What am I trying to do? Get help. I'm trying to make a plan and get help. What am I doing? I have to come up with a plan to get help. Okay. I have to get help.*

My information processing for normal access to my brain's information, prior to this morning's episode, went something like this: I visualize myself sitting in the middle of my brain, which is completely lined with filing cabinets. When I am looking for a thought or an idea or a memory, I scan the

cabinets and identify the correct drawer. Once I find the appropriate file, I then have access to all of the information in
that file. If I don't immediately find what I'm looking for, then
I put my brain back on scan and eventually I access the right
data.

But this morning, my information processing was completely aberrant. Even though my brain remained lined with
filing cabinets, it was as if all the drawers had been slammed
shut and the cabinets pushed just beyond my reach. I was
aware that I knew all this stuff, that my brain held a wealth of
information. But where was it? If the information was still
there, I could no longer retrieve it. I wondered if I would ever
reconnect with linguistic thought or retrieve the mental images of my life. I was saddened that perhaps those portions
of my mind were now lost forever.

Devoid of language and linear processing, I felt discon-
nected from the life I had lived, and in the absence of my
cognitive pictures and expansive ideas, time escaped me. The
memories from my past were no longer available for recollection, leaving me cloaked from the bigger picture of who I was
and what I was doing here as a life form. Focused completely
in the present moment, my pulsing brain felt like it was
gripped in a vice. And here, deep within the absence of
earthly temporality, the boundaries of my earthly body dis-
solved and I melted into the universe.

As the hemorrhaging blood interrupted the normal functioning of my left mind, my perception was released from its
attachment to categorization and detail. As the dominating fibers of my left hemisphere shut down, they no longer inhibited my right hemisphere, and my perception was free to shift
such that my consciousness could embody the tranquility of
my right mind. Swathed in an enfolding sense of liberation
and transformation, the essence of my consciousness shifted

into a state that felt amazingly similar to my experience in Thetaville. I'm no authority, but I think the Buddhists would say I entered the mode of existence they call Nirvana.

In the absence of my left hemisphere's analytical judgment, I was completely entranced by the feelings of tranquility, safety, blessedness, euphoria, and omniscience. A piece of me yearned to be released completely from the captivity of this physical form, which throbbed with pain. But providentially, in spite of the attraction of this unremitting temptation, something inside of me remained committed to the task of orchestrating my rescue, and it persevered to ultimately save my life.

Stumbling into my office space, I turned the lights down low because the light stimulation burned my brain like wildfire. The harder I tried to stay focused and concentrate on what I was doing in the here and now, the more intense the throbbing in my head reverberated. It took great effort just to stay attentive and my mind groped to hold on, to remember, *What is it I'm doing? What am I doing? Call for help, I'm trying to call for help!* I vacillated between moments of being able to think clearly (I call these "waves of clarity"), and the lack of ability to think at all.

Feeling cast out of synchrony with the life I had known, I was concurrently disturbed and fascinated by what I was witnessing as the systematic breakdown of my cognitive mind. Time stood still because that clock that would sit and tick in the back of my left brain, that clock that helped me establish linearity between my thoughts, was now silent. Without the internal concept of relativity or the complementary brain activity that helped me navigate myself linearly, I found myself floating from isolated moment to isolated moment. "A" no longer had any relationship to "B" and "one" was no longer relative to "two." These types of sequences required an intellectual connection that my mind could no longer perform. Even the simplest of calculations, by definition, requires recognition of the relationship between different entities, and my mind was no longer capable of creating combinations. So again, I sat befuddled, waiting for the next intermittent thought or wave of clarity. In anticipation of the eventual arrival of an idea that would connect me to something in objective reality, my mind kept repeating, *What is it I'm trying to do?*

Why didn't I just call 9-1-1? The hemorrhage growing in my cranium was positioned directly over the portion of my left brain that understood what a number was. The neurons that coded 9-1-1 were now swimming in a pool of blood, so the concept simply didn't exist for me anymore. Why didn't I just skip downstairs and ask my landlady for help? She was home on maternity leave and would have been happy to give me a lift. But her file, again, a detail in the big picture of my life in relation to those around me, didn't exist anymore. Why didn't I walk out into the street and flag down a stranger for help? It never crossed my mind. In this incapacitation, the only option I had was the one I was desperate to remember—which was how to call for help!

All I could do was sit and wait; sit patiently with the phone by my side and wait in the silence. So there I sat, home alone with these transient thoughts that evaded me, almost teasing me as they flitted in and out of my mind. I sat waiting for a wave of clarity that would permit my mind to connect two thoughts and give me a chance at forming an idea, a chance to execute a plan. I sat silently intoning, *What am I doing? Call for help. Call for help. I'm trying to call for help.*

In the hope that I might consciously evoke another wave of clarity, I placed the phone on the desk in front of me and stared at its keypad. Searching for some recollection of a number to dial, my wandering brain felt empty and sore as I forced it to concentrate and pay attention. Pulsing, pulsing, pulsing. Gosh my brain hurt. In an instant a number flashed through my mind's eye. It was my mother's number. How thrilling that I could remember! How wonderful that I could not only recall a number but that I knew whose number it was! And how remarkable, though unfortunate, that even in this precarious condition, I realized that my mother lived more than a thousand miles away and how inappropriate it would be to call her now. I thought to myself, *No way, I can't call Mama and tell her I'm having a stroke! That would be horrible. She would freak out! I've got to come up with a plan.*

In a moment of clarity, I knew that if I called work, my colleagues at the Harvard Brain Tissue Resource Center would get me help. *If only I could remember the number at work.* And how ironic it was that I had spent the previous two years singing the Brain Bank jingle to audiences all around the country, including the lyrics, "Just dial 1-800-BrainBank for information please!" But on this morning, with all those memories set beyond my reach, I retained only a vague idea of who I was and what I was trying to accomplish. Posed at my desk in a bizarre mental fog, I continued to coax my mind by obsessing, *What is the number at work? Where do I work? The Brain Bank. I work at the Brain Bank. What is the number at the Brain Bank? What am I doing? I'm calling for help. I'm calling work. Okay, what is the number at work?*

My normal perception of this external world had been successfully established by the constant exchange of information between my right and left hemispheres. Because of cortical laterality, each half of my brain specialized in slightly varied functions, and when put together, my brain could precisely manufacture a realistic perception of the external world. Although I had been a very bright child with tremendous potential for learning, my two hemispheres had never been equal in their natural abilities. My right hemisphere excelled at understanding the big picture of ideas and concepts, but my left hemisphere had to work extremely hard to memorize random facts and details. As a result, I was one of those people who rarely chose to cognitively code a phone number as a random sequence of numbers. Instead, my mind automatically created some sort of pattern, most often a visual pattern, to which I attached the sequence. In the case of phone numbers, I generally memorized the pattern as it dialed on a touch-tone keypad. Privately, I always wondered how I would have survived in a world of rotary telephones where such schematic ploys would have been much more challenging!

Throughout my youth, my mind had been much more interested in how things were intuitively related (right hemisphere) than how they were categorically different (left hemisphere). My mind preferred thinking in pictures (right hemisphere), as opposed to language (left hemisphere). It wasn't until my graduate school years and fascination with anatomy that my mind excelled in detail memorization and retrieval. After a childhood of information processing through sensory, visual, and pattern association strategies, the tapestry of my knowledge was all intimately interlinked.

The downfall to this type of a learning system, of course, is that it only works when all the pieces of the network are functioning and interacting properly. On this morning, as I sat and contemplated the phone number for work, I remembered that there was something unique about the patterning of our office codes. Something like, my number ended in 1-0; which was the exact opposite of my boss's number which ended in 0-1; and my colleague's number fell right in the middle. But because my left hemisphere was drowning in a puddle of blood, I could not access the specifics of my mental inquiry, and the linearity of mathematics befuddled me. I kept think-

ing, *What's in the middle between 01 and 10?* I decided that
looking at the phone keypad might be helpful.

Sitting at my desk, I placed the phone directly in front of
me and sat patiently for a few moments awaiting the next
wave of clarity. Again I intoned, *What is the number at work?
What is the number at work?* After several minutes of holding
the phone and drawing a blank, a list of four digits suddenly
appeared in my mind . . . 2405! 2405! I repeated it over and
over to myself . . . *2405!* In order to not forget it, I picked up
a pen and with my nondominant left hand, I quickly jotted
down the image I saw in my mind. A "2" was no longer a "2"
but rather a squiggle that looked like a "2." Fortunately, the
"2" on the phone pad looked just like the "2" in my mind's
eye, so I drew the squiggles that represented what I saw . . .
2405. Somehow I understood that this was only part of the
number, what was the rest? There was a prefix—something
came first. So, again, I started intoning, *What is the prefix?
What is the prefix at work?*

Faced with this dilemma, it occurred to me that it was not
necessarily an advantage that, when we are at work, we
merely have to dial extension numbers. Because of this lack
of routine use, the pattern for my prefix recognition was not
coded in the exact same file in my brain as the rest of the
extension numbers. So back I went on a mission to retrieve
information and I questioned, *What is the prefix? What is the
prefix at work?*

For my entire life, I had been exposed to phone numbers
with very low prefixes: 232, 234, 332, 335, etc. But grasping at
anything flitting through my mind, any possibility at all, the
code 855 flashed as a visual. Initially, I thought that this was
the most absurd prefix I had ever heard, because the numbers seemed so high. But at this point, anything was worth a
try. In anticipation of the next wave of clarity, I cleared the

desk in front of me. Because it was only 9:15 A.M., and I was only fifteen minutes late for work, no one would really be missing me yet. With a plan in mind, I plodded on.

I felt tired. I felt vulnerable and completely fragmented as I sat there waiting. Although I was consistently distracted by an enveloping sense of being at *one* with the universe, I was desperate to carry out my plan to get help. Within my mind, I rehearsed over and over again what I needed to do, and what I would say. But keeping my mind tuned in to what I was trying to do was like struggling to hang on to a slippery fish. Task one, hold the thought in mind; task two, execute the internal perception in the external world. Pay attention. Hold on to the fish. Hold on to the understanding that this is a phone. Hold on. Hold on for the next functional moment of clarity! I kept rehearsing in my mind, *This is Jill. I need help! This is Jill. I need help!*

This process had already taken forty-five minutes for me to figure out who and how to call for help. During the next wave of clarity, I dialed the number by matching the squiggles on the paper to the squiggles on the phone pad. To my great fortune, my colleague and good friend, Dr. Stephen Vincent, was sitting at his desk. As he picked up the receiver, I could hear him speak, but my mind could not decipher his words. I thought, *Oh my gosh, he sounds like a golden retriever!* I realized that my left hemisphere was so garbled that I could no longer understand speech. Yet, I was so relieved to be connected to another human being that I blurted out, "This is Jill. I need help!" Well, at least that's what I tried to say. What exactly came out of my mouth was more akin to grunts and groans, but fortunately Steve recognized my voice. It was clear to him that I was in some sort of trouble. (Apparently all those years of hollering up and down the halls at work had earned me a recognizable squawk!)

I was shocked, however, when I did realize that I could not speak intelligibly. Even though I could hear myself speak clearly within my mind—*This is Jill, I need help!*—the sounds coming out of my throat did not match the words in my brain. I was disturbed to comprehend that my left hemisphere was even more disabled than I had realized. Although my left hemisphere could not decipher the meaning of the words he spoke, my right hemisphere interpreted the soft tones in his voice to mean that he would get me help.

Territory of Jill's Hemorrhage
(shaded oval area)

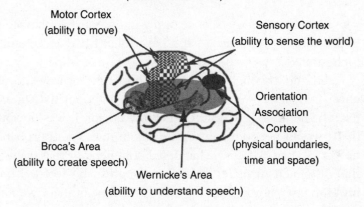

Motor Cortex
(ability to move)

Sensory Cortex
(ability to sense the world)

Orientation
Association
Cortex
(physical boundaries,
time and space)

Broca's Area
(ability to create speech)

Wernicke's Area
(ability to understand speech)

Finally, in that moment, I could relax. I didn't need to understand the details of what he would do. I knew that I had done all that I could do; all that anyone could have hoped that I would do, to save myself.

SIX

My Return to the Still

A s I sat there in the silence of my mind, satisfied that Steve would get me help, I felt relieved that I had successfully orchestrated my rescue. My paralyzed arm was partially re-covered and although it hurt, I felt hopeful that it would re-cover completely. Yet even in this discombobulated state, I felt a nagging obligation to contact my doctor. It was obvious that I would need emergency treatment that would probably be very expensive, and what a sad commentary that even in this disjointed mentality, I knew enough to be worried that my HMO might not cover my costs in the event that I went to the *wrong* health center for care.

Still sitting at my desk, with my good left arm I reached for the three-inch stack of business cards I had collected over the past few years. I had only visited my current doctor once, about six months earlier, but I remembered that there was something Irish about her name—St. something, St. some-thing, so I began searching for associations. In my mind's eye, I could recall perfectly the symbol of the Harvard crest located in the top central position of her business card. Pleased with my ability to remember exactly what the card looked like, I thought to myself, *Fine, this will be just fine; all I have to do is find the card and make the call.*

To my astonishment, however, as I looked at the top card, I realized that although I retained a clear image in my mind of what I was looking for, I could not discriminate any of the information on the card in front of me. My brain could no longer distinguish writing as writing, or symbols as symbols, or even background as background. Instead, the card looked like an abstract tapestry of pixels. The entire picture was a uniform blend of all its constituent pieces. The dots that formed the symbols of language blended in smoothly with the dots of the background. The distinctions of color and edge no longer registered to my brain.

Dismayed, I realized that my ability to interact with the external world had deteriorated far more than I could ever have imagined. My grip on normal reality had been all but peeled away. I was no longer capable of perceiving the mental cues I had depended on to visually discriminate between objects. On top of my inability to identify my own physical boundaries, and in the absence of my internal clock, I perceived myself as fluid. Coupled with my loss of long-term and short-term memories, I no longer felt grounded or safe in the external world.

What a daunting task it was to simply sit there in the center of my silent mind, holding that stack of cards and trying to remember, *Who am I? What am I doing?* Searching for any connection with my external reality, I had lost all sense of urgency. Yet amazingly, my frontal lobe fought hard to hang on to the task and I still embraced the occasional wave of clarity that routed me back into this earthly realm via my physical pain. During these moments of clarity, I could see, I could identify, I could remember what I was doing, and I could discriminate again between the varied incoming stimuli. So faithfully I plodded forward. *That's not the card, that's not the card, that's not the card.* It took more than thirty-five minutes for me to navigate my way a mere inch down into that stack where I finally recognized the Harvard crest.

By this point, however, the entire concept of a telephone was a very interesting and bizarre kind of thing. I felt oddly removed from my ability to have any comprehension about what it was I was supposed to do with it. Somehow I understood that this "thing" in my space was going to connect me through a wire to a completely different space. And at the other end of the wire, there would be a person to whom I could speak and she would understand me. Wow, imagine that!

Because I was afraid that I would lose my focus and the doctor's card would get confused with the others, I cleared the desk space in front of me and placed her card directly in front. I picked up the phone and placed the number keypad on the desk right next to the business card. Because my brain had been on a steady rate of disintegration, the appearance of the number pad now looked completely strange and foreign. As I sat there drifting in and out of my insubordinate left mind, I remained calm. Periodically, I was able to match the number squiggle on the card to the number squiggle on the telephone keypad. To keep track of the numbers that I had already dialed, I covered the number on the business card with my left index finger as soon as I pushed the number on the phone using my stumpy right index finger. I had to do this because I could not remember from moment to moment which numbers I had already pressed. I repeated this strategy until all the numbers were dialed and then I placed the phone to my ear and listened.

Feeling drained and disoriented, I was afraid that I would forget what I was doing, so I continued to repeat in my mind,

This is Jill Taylor. I'm having a stroke. This is Jill Taylor. I'm having a stroke. But when the phone was answered and I tried to speak, I was blown away to discover that although I could hear myself speaking clearly, within my mind, no sound came out of my throat. Not even the grunts that I was able to produce earlier. I was flabbergasted. *Oh my gosh! I can't talk, I can't talk!* And it wasn't until this moment when I tried to speak out loud that I had any idea that I couldn't. My vocal cords were inoperative and nothing, no sound at all, would come forth.

Like priming a pump, I pushed air forcefully out of my chest and inhaled deeply, over and over again, trying to make some sound, trying to make any sound come out. Realizing what I was doing, I thought, *They're going to think this is an obscene phone call! Don't hang up! Please don't hang up!* But just like priming a pump, repeatedly pushing the air in and out, forcing my chest and my throat to vibrate, "Uhhhhhh, uhhhhhh, thhhhhh, thhhhhhe, thhhhhiiiiiiiizzzxzzaaaaaaa" finally came out. The call was immediately forwarded from the receptionist to my doctor, who miraculously just happened to be sitting in for office hours! With the patience of a gentle soul, she sat and she listened as I struggled to enunciate, "This is Jill Taylor. I'm having a stroke."

Eventually my doctor understood enough of my message to comprehend who I was and what I needed. She directed me, "Get to Mount Auburn Hospital." As she spoke, although I could hear her words, I could not grasp their meaning. Feeling despondent, I thought to myself, *If only she would speak more slowly and enunciate more clearly, perhaps I could get it, perhaps I could understand.* With hope in my heart, I pleaded in a semi-intelligible way, "Again?" With concern, she slowly repeated her directive, "Get to Mount Auburn Hospital." Yet again, I could not comprehend. With patience and genuine compassion for my obvious neurological breakdown, she repeated her directive. Repeatedly, I could not connect meaning to the sounds and make sense of what she said. Feeling exasperated by my own inability to understand her simple language, I primed my vocal pump again and somehow communicated that help was on its way and we would call her back.

At this point, it didn't take a brain scientist to understand what was going on in my brain. The longer the blood from the hemorrhage continued to spill into my cortex, the more massive the tissue damage would become and the more cognitively inept I would be. Although the AVM originally burst near the middle to posterior portion of my cerebral cortex in my left hemisphere, by this point, the cells in my left frontal lobe—responsible for my ability to generate language, were also compromised. It was predictable that as the blood interrupted the flow of information transmission between my two language centers (Broca's anteriorly and Wernicke's posteriorly, page 40), I could neither create/express language nor understand it. At this point in time, however, my greatest concern was that my vocal cords were not responding to my mental cues. I still feared that the centers in the pons of my brainstem, including my center for inspiration, were possibly at risk.

Feeling defeated and tired, I hung up the phone. Rising from my seat, I wrapped a scarf around my head to block the streaming light from my eyes. Picturing the deadbolt on my front door, I slowly navigated my body, step by step, down the front flight of stairs by sliding on my butt. Ready for company and no longer preoccupied with what I felt compelled to do, I crawled back up the stairs to my living room, where I crouched on my couch to quiet my weary mind.

Despondent and alone, I felt discomfort in my pulsating

head, and I communed with my wound as I acknowledged the degeneration of my connection to this life. With every moment that passed, I felt my connection with my body becoming weaker. I sensed that my energy was leaking out of this fragile container—deadening the distal tips of my fingers and toes. I could hear the machinery of my body grinding its wheels as my cells systematically manufactured my life, and I feared that my cognitive mind was becoming so disabled, so detached from its normal ability to function, that I would be rendered permanently disabled. For the first time in my life, I understood that I was not invincible. Unlike a computer that could be turned off and then rebooted, the richness of my life was completely dependent on not only the health of my cellular structure, but on the integrity of my brain's ability to electrically transmit and communicate its directives.

Humbled by the direness of my situation, I grieved for the loss of my life as I anticipated the death and degeneration of my cellular matrix. Despite the overwhelming presence of the engulfing bliss of my right mind, I fought desperately to hold on to whatever conscious connections I still retained in my left mind. By now, I understood clearly that I was no longer a normal human being. My consciousness no longer retained the discriminatory functions of my dominant analytical left brain. Without those inhibiting thoughts, I had stepped beyond my perception of myself as an individual. Without my left brain available to help me identify myself as a complex organism made up of multiple interdependent systems or to define me as a distinct collection of fragmented functions, my consciousness ventured unfettered into the peaceful bliss of my divine right mind.

As I sat in the silence and pondered my new perceptions, I wondered how disabled I could become before the loss would be permanent. I contemplated how many circuits I could lose and how detached from my higher cognitive abilities I could tread and still have any hope of ever regaining normal function. I hadn't come this far to just die or become mentally vegetative! So I held my head in my hands and wept. Amidst my tears, I clenched my fists and prayed. I prayed for peace in my heart. I prayed for peace in my mind and I prayed, *Please, Great Spirit, don't shut down my life.* And into the silence my mind implored, *Hold on. Be still. Be quiet. Hold on.*

I sat there in the middle of my living room for what seemed to be an eternity. When Steve appeared in the doorway, no words were exchanged. I handed him my doctor's card and he immediately called for instructions. Promptly, he escorted me down the stairs and out the door. Gently, he guided me to his car, strapped me in, and reclined the seat. He wrapped my head with a scarf to shade my eyes from the light. He spoke softly, encouragingly patted my knee, and proceeded to drive to Mount Auburn Hospital.

By the time we arrived, I was still conscious but obviously delirious. They placed me in a wheelchair and led us into the waiting room. Steve was clearly distressed with their indifference to the severity of my condition, but he obediently filled out my paperwork and helped me sign my name. Waiting for our turn, I felt the energy in my body shift and like a balloon I deflated into my own lap, shifting into a semiconscious condition. Steve insisted that I receive immediate attention!

I was taken to have a CT scan of my brain. They lifted me out of the wheelchair and placed me on the CT gurney. Despite the throbbing pain in my brain that was echoed by the thumping sounds of the machine's motor, I was conscious enough to find some satisfaction in learning that my self-diagnosis had been correct. I was experiencing a rare form of stroke. I had a massive hemorrhage flooding the left hemisphere of my brain. Although I don't recall it, my medical records indicate that I was given an initial dose of steroids to slow the inflammation.

The agenda was to ship me immediately to Massachusetts General Hospital. My gurney was lifted and bolted into position in an ambulance for the ride across Boston. I remember that a kindhearted paramedic accompanied me along my journey. With compassion, he wrapped me in a blanket and arranged a jacket over my face to protect my eyes. His touch upon my back was comforting; his gentle kindness, priceless.

I was finally free from worry. I curled up into a fetal ball and lay waiting. I understood that, on this morning, I had witnessed the step-by-step deterioration of my intricate neurological circuitry. I had always celebrated my life as a magnificent physical manifestation of my DNA, and oh, what a colorful genetic pool from which I had been spawned! For thirty-seven years, I had been blessed with an agile mosaic of electrified biochemistry. And, like many folks, I had fantasized that I wanted to be awake when I died because I wanted to witness that remarkable final transition.

Just before noon, on December 10, 1996, the electrical vitality of my molecular mass grew dim, and when I felt my energy lift, my cognitive mind surrendered its connection to, and command over, my body's physical mechanics. Sanctioned deep within a sacred cocoon with a silent mind and a tranquil heart, I felt the enormousness of my energy lift. My body fell limp, and my consciousness rose to a slower vibration. I clearly understood that I was no longer the choreographer of this life. In the absence of sight, sound, touch, smell, taste, and fear, I felt my spirit surrender its attachment to this body and I was released from the pain.

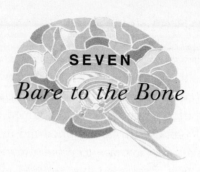

SEVEN

Bare to the Bone

Upon my arrival at the Massachusetts General Hospital emergency ward, I landed in the center of an energy spin that I could only describe as a bustling beehive. My limp body felt heavy and terribly weak. It was drained of all its energy—like a balloon that had been slowly and thoroughly deflated. Medical personnel swarmed about my gurney. The sharp lights and intense sounds beat upon my brain like a mob, demanding more attention than I could possibly muster to appease them.

"Answer this, squeeze that, sign here!" they demanded of my semiconsciousness, and I thought, *How absurd! Can't you see I've got a problem here? What's the matter with you people? Slow down! I can't understand you! Be patient! Hold still! That hurts! What is this chaos?* The more tenaciously they tried to draw me out, the greater was my ache to reach inside for my personal source of sustenance. I felt besieged by their touching, probing, and piercing; like a slug sprinkled with salt, I writhed in response. I wanted to scream *Leave me alone!* but my voice had fallen silent. They couldn't hear me because they couldn't read my mind. I passed out like a wounded animal, desperate to escape their manipulations.

When I first awoke later that afternoon, I was shocked to discover that I was still alive. (Heartfelt thanks to the medical professionals who stabilized my body and gave me another chance at life—even though no one had any idea what or how much I would ever recover.) My body was draped in the customary hospital gown and I was resting in a private cubicle. The bed was partially raised with my aching head slightly elevated on a pillow. Devoid of my usual well of energy, my body sank deep into the bed like a lump of heavy metal that I couldn't begin to budge. I could not determine how my body was positioned, where it began or where it ended. Without the traditional sense of my physical boundaries, I felt that I was at *one* with the vastness of the universe.

My head pulsed with a tormenting pain that pounded like thunder while a white lightning storm raged theatrically upon the inside of my eyelids. Every little shift in position that I tried to make required more energy than my reserves held. Simply inhaling hurt my ribs, and the light that flooded in through my eyes burned my brain like fire. Not able to speak, I begged for the lights to be dimmed by burying my face into a sheet.

I couldn't hear anything beyond the pounding rhythm of my heart, which pulsed so loudly that my bones vibrated with ache and my muscles twitched with anguish. My keen scientific mind was no longer available to record, relate, detail, and categorize information about the surrounding external three-dimensional space. I wanted to wail like a colicky newborn that had been suddenly plunged into a realm of chaotic stimulation. With my mind stripped of its ability to recall the memories and details of my previous life, it was clear to me that I was now like an infant—born into an adult woman's body. And oh yes, the brain wasn't working!

Here in this emergency cubby, I could sense, over my left shoulder, the presence of two familiar associates as they peered at a CT scan mounted on a wall light-box. The picture on display contained serial sections of my brain, and although I could not decipher the words my colleagues softly shared, their body language communicated the gravity of the situation. It didn't take someone with a Ph.D. in neuroanatomy to figure out that the huge white hole in the middle of the brain scan didn't belong there! My left hemisphere was swimming in a pool of blood and my entire brain was swollen in response to the trauma.

In silent prayer, I reflected, *I am not supposed to be here anymore! I let go! My energy shifted and the essence of my being escaped. This is not right. I don't belong here anymore! Great Spirit,* I mused, *I am now at one with the universe. I have blended into the eternal flow and am beyond returning to this plane of life—yet I remain tethered here. The fragile mind of this organic container has shut down and is no longer amenable for intelligent occupancy! I don't belong here anymore!* Unencumbered by any emotional connection to anyone or anything outside of myself, my spirit was free to catch a wave in the river of blissful flow. *Let me out!* I hollered within my mind, *I let go! I let go!* I wanted to escape this vessel of physical form, which radiated chaos and pain. In those brief moments, I felt tremendous despair that I had survived.

My body felt cold, weighty, and ached with pain. The signals between my brain and body were so defective that I couldn't recognize my physical form. I felt as if I was an electrical being; an apparition of energy smoldering around an organic lump. I had become a pile of waste, leftovers, but I still retained a consciousness. A consciousness that was different from the one I had known before, however, because my left hemisphere had been packed with details about how to make sense of the external world. These details had been organized and ingrained as neuronal circuits in my brain. Here, in the absence of that circuitry, I felt inanimate and awkward. My consciousness had shifted. I was still in here—I was still me, but without the richness of the emotional and cognitive connections my life had known. So, was I really still me? How could I still be Dr. Jill Bolte Taylor, when I no longer shared her life experiences, thoughts, and emotional attachments?

I remember that first day of the stroke with terrific bittersweetness. In the absence of the normal functioning of my left orientation association area, my perception of my physical boundaries was no longer limited to where my skin met air. I felt like a genie liberated from its bottle. The energy of my spirit seemed to flow like a great whale gliding through a sea of silent euphoria. Finer than the finest of pleasures we can experience as physical beings, this absence of physical boundary was one of glorious bliss. As my consciousness dwelled in a flow of sweet tranquility, it was obvious to me that I would never be able to squeeze the enormousness of my spirit back inside this tiny cellular matrix.

My escape into bliss was a magnificent alternative to the daunting sense of mourning and devastation I felt every time I was coaxed back into some type of interaction with the percolating world outside of me. I existed in some remote space that seemed to be far away from my normal information processing, and it was clear that the "I" whom I had grown up to be had not survived this neurological catastrophe. I understood that that Dr. Jill Bolte Taylor died that morning, and yet, with that said, who was left? Or, with my left hemisphere destroyed, perhaps I should now say, who was *right?*

Without a language center telling me: "I am Dr. Jill Bolte Taylor. I am a neuroanatomist. I live at this address and can be reached at this phone number," I felt no obligation to being her anymore. It was truly a bizarre shift in perception, but without her emotional circuitry reminding me of her likes and dislikes, or her ego center reminding me about her patterns of critical judgment, I didn't think like her anymore. From a practical perspective, considering the amount of biological damage, being her again wasn't even an option! In my mind, in my new perspective, that Dr. Jill Bolte Taylor died that morning and no longer existed. Now that I didn't know her life—her relationships, successes, and mistakes, I was no longer bound to her decisions or self-induced limitations.

Although I experienced enormous grief for the death of my left hemisphere consciousness—and the woman I had been, I concurrently felt tremendous relief. That Dr. Jill Bolte Taylor had grown up with lots of anger and a lifetime of emotional baggage that must have required a lot of energy to sustain. She was passionate about her work and her advocacy. She was intensely committed to living a dynamic life. But despite her likable and perhaps even admirable characteristics, in my present form I had not inherited her fundamental hostility. I had forgotten about my brother and his illness. I had forgotten about my parents and their divorce. I had forgotten about my job and all the things in my life that brought me stress—and with this obliteration of memories, I felt both relief and joy. I had spent a lifetime of thirty-seven years being enthusiastically committed to "do-do-doing" lots of stuff at a very fast pace. On this special day, I learned the meaning of simply "being."

When I lost my left hemisphere and its language centers, I

also lost the clock that would break my moments into consecutive brief instances. Instead of having my moments prematurely stunted, they became open-ended, and I felt no rush to do anything. Like walking along the beach, or just hanging out in the beauty of nature, I shifted from the doing-consciousness of my left brain to the being-consciousness of my right brain. I morphed from feeling small and isolated to feeling enormous and expansive. I stopped thinking in language and shifted to taking new pictures of what was going on in the present moment. I was not capable of deliberating about past or future-related ideas because those cells were incapacitated. All I could perceive was right here, right now, and it was beautiful.

My entire self-concept shifted as I no longer perceived myself as a single, a solid, an entity with boundaries that separated me from the entities around me. I understood that at the most elementary level, I am a fluid. Of course I am a fluid! Everything around us, about us, among us, within us, and between us is made up of atoms and molecules vibrating in space. Although the ego center of our language center prefers defining our *self* as individual and solid, most of us are aware that we are made up of trillions of cells, gallons of water, and ultimately everything about us exists in a constant and dynamic state of activity. My left hemisphere had been trained to perceive myself as a solid, separate from others. Now, released from that restrictive circuitry, my right hemisphere relished in its attachment to the eternal flow. I was no longer isolated and alone. My soul was as big as the universe and frolicked with glee in a boundless sea.

For many of us, thinking about ourselves as fluid, or with souls as big as the universe, connected to the energy flow of all that is, slips us out just beyond our comfort zone. But without the judgment of my left brain saying that I am a solid, my perception of myself returned to this natural state of fluidity. Clearly, we are each trillions upon trillions of particles in soft vibration. We exist as fluid-filled sacs in a fluid world where everything exists in motion. Different entities are composed of different densities of molecules but ultimately every pixel is made up of electrons, protons, and neutrons performing a delicate dance. Every pixel, including every iota of you and me, and every pixel of space seemingly in between, is atomic matter and energy. My eyes could no longer perceive things as things that were separate from one another. Instead, the energy of everything blended together. My visual processing was no longer normal. (I compare this pixilated perspective to Impressionist pointillism paintings.)

I was consciously alert and my perception was that I was in the flow. Everything in my visual world blended together, and with every pixel radiating energy we all flowed *en masse*, together as *one*. It was impossible for me to distinguish the physical boundaries between objects because everything radiated with similar energy. It's probably comparable to when people take off their glasses or put eye drops into their eyes— the edges become softer.

In this state of mind, I could not perceive three-dimensionally. Nothing stood out as being closer or farther away. If there was a person standing in a doorway, I could not distinguish their presence until they moved. It took activity for me to know that I should pay special attention to any particular patch of molecules. In addition, color did not register to my brain as color. I simply couldn't distinguish it.

Prior to this morning, when I had experienced myself as a solid, I had possessed the ability to experience loss—either physical loss via death or injury, or emotional loss through heartache. But in this shifted perception, it was impossible for me to perceive either physical or emotional loss because I was not capable of experiencing separation or individuality. Despite my neurological trauma, an unforgettable sense of peace pervaded my entire being and I felt calm.

Although I rejoiced in my perception of connection to all that is, I shuddered at the awareness that I was no longer a normal human being. How on earth would I exist as a member of the human race with this heightened perception that we are each a part of it all, and that the life force energy within each of us contains the power of the universe? How could I fit in with our society when I walk the earth with no fear? I was, by anyone's standard, no longer normal. In my own unique way, I had become severely mentally ill. And I must say, there was both freedom and challenge for me in recognizing that our perception of the external world, and our relationship to it, is a product of our neurological circuitry. For all those years of my life, I really had been a figment of my own imagination!

When the time keeper in my left hemisphere shut down, the natural temporal cadence of my life s-l-o-w-e-d to the pace of a snail. As my perception of time shifted, I fell out of sync with the beehive that bustled around me. My consciousness drifted into a time warp, rendering me incapable of communicating or functioning at either the accustomed or acceptable pace of social exchange. I now existed in a world between worlds. I could no longer relate to people outside of me, and yet my life had not been extinguished. I was not only an oddity to those around me, but on the inside, I was an oddity to myself.

I felt so detached from my ability to move my body with any oomph that I truly believed I would never be able to get this collection of cells to perform again. Wasn't it interesting that although I could not walk or talk, understand language, read or write, or even roll my body over, I knew that I was okay? The now offline intellectual mind of my left hemisphere no longer inhibited my innate awareness that I was the miraculous power of life. I knew I was different now—but never once did my right mind indicate that I was "less than" what I had been before. I was simply a being of light radiating life into the world. Regardless of whether or not I had a body or brain that could connect me to the world of others, I saw myself as a cellular masterpiece. In the absence of my left hemisphere's negative judgment, I perceived myself as perfect, whole, and beautiful just the way I was.

You may be wondering how it is that I still remember everything that happened. I remind you that although I was mentally disabled, I was not unconscious. Our consciousness is created by numerous programs that are running at the same time. Each program adds a new dimension to our ability to perceive things in the three-dimensional world. Although I had lost my left hemisphere consciousness containing my ego center and ability to see my *self* as a single and solid entity separate from you, I retained both the consciousness of my right mind and the consciousness of the cells making up my body. Although one set of programs was no longer functioning—the one that reminded me moment by moment of who I was and where I lived, etc., the other parts of me remained alert and continued processing instantaneous information. In the absence of my traditional left hemispheric domination over my right mind, other parts of my brain emerged. Programs that had been inhibited were now free to run and I was no longer fettered to my previous interpretation of perception. With this shift away from my left hemisphere consciousness and the character I had been, my right hemisphere character emerged with new insight.

To hear others tell the story, however, I was quite a mess

that day. I was like a newborn unable to make sense of the sensory stimulation in the physical space around me. It was obvious that I perceived incoming stimulation as painful. Sound streaming in through my ears blasted my brain senseless so that when people spoke, I could not distinguish their voices from the underlying clatter of the environment. From my perspective, everyone clamored *en masse* and resonated like a discordant pack of restless animals. Inside my head, I felt as though my ears were no longer tightly connected to my brain, and I sensed that important information was seeping out between the cracks.

I wanted to communicate: *Yelling louder does not help me understand you any better! Don't be afraid of me. Come closer to me. Bring me your gentle spirit. Speak more slowly. Enunciate more clearly. Again! Please, try again! S-l-o-w down. Be kind to me. Be a safe place for me. See that I am a wounded animal, not a stupid animal. I am vulnerable and confused. Whatever my age, whatever my credentials, reach for me. Respect me. I am in here. Come find me.*

Earlier that morning, I never entertained the possibility that I was orchestrating my rescue so that I would live out the rest of my days completely disabled. Yet, at the core of my being, my conscious mind felt so detached from my physical body that I sincerely believed I would never be able to fit the energy of me back inside this skin, nor ever be able to reengage the intricate networks of my body's cellular and molecular tapestry. I felt suspended between two worlds, caught between two perfectly opposite planes of reality. For me, hell existed inside the pain of this wounded body as it failed miserably in any attempt to interact with the external world, while heaven existed in a consciousness that soared in eternal bliss. And yet, somewhere deep within me, there was a jubilant being, thrilled that I had survived!

EIGHT

Neurological Intensive Care

Once my physicians were satisfied that I was no longer a medical emergency, they moved me up to the Neurological Intensive Care Unit. All I knew was that I had a roommate to my right, my feet faced the door, and my left side was close to a wall. Past that, I didn't have much awareness except for my head and right arm, which both continued to ache.

I experienced people as concentrated packages of energy. Doctors and nurses were massive conglomerations of powerful beams of energy that came and went. I felt rushed by an outer world that did not understand how to communicate with me. Because I could not speak or understand language, I sat silently on the sideline of life. I wish I had a dollar for every time I was given a neurological exam in that first forty-eight hours. Folks buzzed in, probed, prodded, and repeatedly sought neurological information. My energy was drained by these ongoing activities. I would have appreciated it if they would have consolidated their efforts and shared the information.

With this shift into my right hemisphere, I became empathic to what others felt. Although I could not understand the words they spoke, I could read volumes from their facial expression and body language. I paid very close attention to how energy dynamics affected me. I realized that some people brought me energy while others took it away. One nurse was very attentive to my needs: Was I warm enough? Did I need water? Was I in pain? Naturally, I felt safe in her care. She made eye contact and was clearly providing me with a healing space. A different nurse, who never made eye contact, shuffled her feet as though she were in pain. This woman brought me a tray with milk and jello, but neglected to realize that my hands and fingers could not open the containers. I desperately wanted to consume something, but she was oblivious to my needs. She raised her voice when she spoke to me, not realizing that I wasn't deaf. Under the circumstances, her lack of willingness to connect with me scared me. I did not feel safe in her care.

Dr. David Greer was a kind and gentle young man. He was genuinely sympathetic to my situation and took the time to pause during his busy routine to lean down near my face and speak softly to me. He touched my arm to reassure me that I would be okay. Although I could not understand his words, it was clear to me that Dr. Greer was watching over me. He understood that I was not stupid but that I was impaired. He treated me with respect. I'll always be grateful for his kindness.

On that first day, my condition progressed and improved rapidly in some areas, but not at all in others. Although recovery would take years, certain parts of my brain were still intact and eagerly engaged in trying to decipher the billions of bits of data making up the present moment. The most notable difference between my pre- and post-stroke cognitive experience was the dramatic silence that had taken up residency inside my head. It wasn't that I could not think anymore, I just didn't think in the same way. Communication with the external world was out. Language with linear processing was out. But thinking in pictures was in. Gathering glimpses of information, moment by moment, and then taking time to ponder the experience, was in.

One of my doctors asked me the question, "Who is the President of the United States?" In order for me to process this question and come up with an answer, I had to first realize that a question was being asked of me. Once I realized someone wanted my attention, I needed them to repeat the question so I could focus on the sounds being spoken, and then I had to pay really close attention to the movement of their lips. Because it was very difficult for my ears to distinguish a single voice from background noise, I needed the question to be repeated slowly and enunciated clearly. I needed calm, clear communication. I may have had a dense expression on my face and appeared ignorant, but my mind was very busy concentrating on the acquisition of new information. My responses came slowly. Much too slowly for the real world.

Paying attention to what someone was saying took an enormous amount of effort, and I found it to be tiring. First, I had to pay attention with my eyes and ears, neither of which were working normally. My brain had to capture the sound and then match that sound up with a specific lip movement. Then, it had to search and see if there was any meaning for those combinations of sounds stored anywhere in my wounded brain. Once I got one word figured out then I had to search for combinations of words, and with an impaired mind, that took hours!

The effort it took for me to pay attention to what someone was saying was like the effort it takes to pay attention to someone who is speaking on a cell phone with a bad connection. You have to work so hard to hear what the person is saying that you may become impatient, frustrated, and hang up the phone. That's the kind of effort it took for me to hear a voice in a noisy background. It took a tremendous amount of willingness and determination on my part, and infinite patience on the part of the speaker.

For my information processing, I took the sounds of the key words and repeated them over and over again in my brain so that I would not forget what they sounded like. Then I would go on a process of exploration to identify a meaning that matched the sound of those words. *President, President, what is a President? What does that mean?* Once I had a concept (picture) of what a President was then I moved on to the sound United States. *United States, United States, what is a United States? What does that mean?* Once I found the file for United States, again, it was a picture in my mind. Then I had to put together the two images—that of a President and that of the United States. But my doctor was not asking me anything that was really about the United States or about a President. He was asking me to identify a specific man, and that was a completely different file. Because my brain could not get from "President" and "United States" to "Bill Clinton," I gave up—but only after hours of probing and exhausting mental gymnastics.

My ability to cognate was erroneously assessed by how quickly I could recall information, rather than by how my mind strategized to recover the information it held. After all the effort I had put into the task of trying to find the answer to the initial question, it ended up that there were too many associations for me to pick through. Since I was thinking in pictures, I had to start with a single image and then expand upon it. I could not start with the general and find the more specific without exploring the billions of possibilities—which was draining. Perhaps if they had asked me a question about Bill Clinton specifically, then I would have found an image of Bill and then been able to expand from there. If they had asked me, "Who is Bill Clinton married to?" then I would have found an image of Bill Clinton, an image of matrimony, and then hopefully an image of Hillary standing by her man. When using pictures to navigate my way back into language, it was impossible to go from a general file to a specific detail.

To someone looking on, I may have been judged as less than what I had been before because I could not process information like a normal person. I was saddened by the inability of the medical community to know how to communicate with someone in my condition. Stroke is the number one disabler in our society and four times more strokes occur in the left hemisphere, impairing language. I think it is vitally important that stroke survivors share and communicate about how each of their brains strategized recovery. In doing so, our medical professionals could be more effective during those initial hours of treatment and assessment. I wanted my doctors to focus on how my brain was working rather than on whether it worked according to their criteria or timetable. I still knew volumes of information and I was simply going to have to figure out how to access it again.

It really was fascinating for me to watch and experience myself during those earliest stages of recovery. Because of my academics, I intellectually conceptualized my body as a compilation of various neurological programs, but it wasn't until this experience with stroke that I really understood that we all have the ability to lose pieces of ourselves one program at a time. I never really pondered what it would be like to lose my mind, more specifically, my left mind. I wish there were a safe way to induce this awareness in people. It might prove to be enlightening.

Imagine, if you will, what it would feel like to have each of your natural faculties systematically peeled away from your consciousness. First, imagine you lose your ability to make sense of sound coming in through your ears. You are not deaf, you simply hear all sound as chaos and noise. Second, remove your ability to see the defined forms of any objects in your space. You are not blind, you simply cannot see three-dimensionally, or identify color. You have no ability to track an object in motion or distinguish clear boundaries between objects. In addition, common smells become so amplified that they overwhelm you, making it difficult for you to catch your breath.

No longer capable of perceiving temperature, vibration, pain, or proprioception (position of your limbs), your aware-ness of your physical boundaries shift. The essence of your energy expands as it blends with the energy around you, and you sense that you are as big as the universe. Those little voices inside your head, reminding you of who you are and where you live, become silent. You lose memory connection to your old emotional *self* and the richness of this moment, right here, right now, captivates your perception. Everything, including the life force you are, radiates pure energy. With childlike curiosity, your heart soars in peace and your mind explores new ways of swimming in a sea of euphoria. Then ask yourself, how motivated would you be to come back to a highly structured routine?

I did a lot of sleeping that afternoon of the stroke—well, as much sleeping as one can do in a hospital! When I was asleep, I could block out the steady stream of energy that bombarded my senses. By closing my eyes, I could close much of my mind. Light was uncomfortable and my brain throbbed in agony when they shined that bright pen-light to check my papillary reflex. The IV in the back of my hand hurt like salt in an open wound, and I craved being unconscious to their physical manipulations. So I escaped by delving back into the sanctuary of my own silent mind . . . well, at least until the next neurological exam.

Behind the scenes, Steve called my mom G.G. (G.G. is my mother's nickname stemming from her maiden name, Gladys Gillman) to tell her about the day's events. G.G. and Steve had known one another for many years from attending the NAMI national annual conventions. They were quite fond of one another. I'm sure that this was a very difficult phone call for both of them. To hear Steve tell the story, he called her and asked her to sit down. He explained that I had experienced a major cerebral hemorrhage in the left hemisphere of my brain, and that I was currently at Massachusetts General Hospital. He assured her that the physicians had stabilized my body and that I was receiving the best care possible.

Later in the day, my boss Francine called G.G. and encouraged her to take a couple of days to get her affairs in order so she could come to Boston for an extended visit. It was clear to Francine that I would probably need surgery. She hoped G.G. would be able to come and give me long-term care in the Boston area. G.G. never hesitated. She spent ten years of her life trying to help my brother heal his mind, to no avail. However, she felt that she could help this child recover from her neurological trauma. G.G. turned all those years of frustration for not being able to heal my brother's schizophrenia into a plan for helping me recover my mind.

NINE

Day Two: The Morning After

I awoke early the next morning to a medical student who came rushing in to take a medical history. I thought it curious that she had not been informed that I was a stroke survivor who could not speak or understand language. I realized that morning that a hospital's number one responsibility should be protecting its patients' energy levels. This young girl was an energy vampire. She wanted to take something from me despite my fragile condition, and she had nothing to give me in return. She was rushing against a clock and obviously losing the race. In her haste, she was rough in the way she handled me and I felt like a detail that had fallen through someone's crack. She spoke a million miles a minute and hollered at me as if I were deaf. I sat and observed her absurdity and ignorance. She was in a hurry and I was a stroke survivor—not a natural match! She might have gotten something more from me had she come to me gently with patience and kindness, but because she insisted that I come to her in her time and at her pace, it was not satisfying for either of us. Her demands were annoying and I felt weary from the encounter.

I realized that I would have to protect my precious energy with keen caution.

The biggest lesson I learned that morning was that when it came to my rehabilitation, I was ultimately the one in control of the success or failure of those caring for me. It was my decision to show up or not. I chose to show up for those professionals who brought me energy by connecting with me, touching me gently and appropriately, making direct eye contact with me, and speaking to me calmly. I responded positively to positive treatment. The professionals who did not connect with me sapped my energy, so I protected myself by ignoring their requests.

Making the decision to recover was a difficult, complicated, and cognitive choice for me. On the one hand, I loved the bliss of drifting in the current of the eternal flow. Who wouldn't? It was beautiful there. My spirit beamed free, enormous, and peaceful. In the rapture of an engulfing bliss, I had to question what recovery really meant. Clearly, there were some advantages to having a functional left hemisphere. It would allow me the skills of interacting with the external world again. In this state of disability, however, attending to what I perceived as chaos was pure pain, and the effort it would take for me to recover, well, was that my priority?

Honestly, there were certain aspects of my new existence that I preferred over the way I had been before. I was not willing to compromise my new insights in the name of recovery. I liked knowing I was a fluid. I loved knowing my spirit was at *one* with the universe and in the flow with everything around me. I found it fascinating to be so tuned in to energy dynamics and body language. But most of all, I loved the feeling of deep inner peace that flooded the core of my very being.

I yearned to be in a place where people were calm and

valued my experience of inner peace. Because of my heightened empathy, I found that I was overly sensitive to feeling other people's stress. If recovery meant that I had to feel like they felt all the time, I wasn't interested. It was easy for me to separate my "stuff" and emotions from other people's "stuff" and emotions by choosing to observe but not engage. As Marianne Williamson puts it, "Could I rejoin the rat race without becoming a rat again?"

Andrew, another medical student, came by that same morning to give me yet another neurological exam. I was wobbly, incredibly weak, and not capable of sitting up by myself, much less capable of standing up on my feet. But because he was gentle yet firm in his touch, I felt safe with him. He spoke calmly, looked me directly in the eyes, and repeated himself as needed. He was respectful of me as a person— even in this condition. I was confident he would grow up to be a fine doctor. I hope that he has.

Dr. Anne Young, who was, at that time, the chairperson of the Massachusetts General Hospital Department of Neurology (I call her the Queen of Neurology), was my neurologist. I had heard about the famous Anne Young for years while working at the Harvard Brain Bank. She served on the Advisory Committee for the Harvard Brain Bank and just two weeks earlier, it was my privilege to sit next to her at an Advisory Counsel luncheon held at the annual Neuroscience Meetings in New Orleans. At the luncheon, I presented the outreach efforts I was engaged in to increase the number of brains donated for research by the psychiatrically diagnosed population. Dr. Young had met the "professional me" that day, so by the time she found me on her morning roster, we had already established a special rapport.

Among the many circuits in my brain that had gone offline, it was my good fortune that my circuitry for embarrassment had also gone awry. Very much like a mother duck followed by her long row of ducklings, Dr. Young and her entourage of medical students arrived at my doorway for morning rounds. To my retrospective horror, I was buck-naked with my derriere in the air and in the middle of a sponge bath, when the Queen of Neurology and her party arrived!

Dr. Young's eyes were soft and kind, and she smiled as she looked me straight in the eye. When she approached, she immediately reached for my foot—much like a good horse handler will touch a horse on their backside as they pass behind it. Dr. Young helped me into a comfortable position. She then stood by my shoulder, gently resting her hands on my arm, and spoke softly to me—not to her students, but to me. She leaned over the edge of my bed and got close enough to my face that I could hear her. Although I could not completely understand her words, I completely understood her intention. This woman understood that I was not stupid but that I was wounded, and it was clear that she knew that it was her job to figure out which circuits of mine were still active and which parts needed healing.

Dr. Young respectfully asked me if it was okay that she teach her students about the neurological exam, and I agreed. As it turned out, I was the brain scientist who failed every task on cue and Dr. Young did not leave my bedside until she was confident that I had no more need of her. On her way out the door, she squeezed my hand and then my toe. I felt a huge sense of relief that she was my physician. I felt that she understood me.

Later that morning, it was time for me to have an angiogram that would outline the blood vessels in my brain. We needed a really good picture of exactly what type of hemorrhage I had had, and the angiogram was the test of choice. Although I thought it absolutely absurd that anyone would ask me to sign a form of consent while in this condition, I realized that policy is policy! How do we define "of sound mind and body" anyway?

Bad news certainly travels fast. Word of my stroke surged through the networks of both McLean Hospital and the membership of NAMI. Here I was, the youngest national board member they had ever elected, having a stroke at thirty-seven.

Two of my colleagues from the Brain Bank came to visit while I was in the Neurology ICU that afternoon. Mark and Pam brought a little stuffed bear for me to cuddle, and I was grateful for their kindness. Although I could sense their initial trepidation, they brought me positive energy and told me, "You're Jill, you're going to be just fine." This confidence in my complete recovery was priceless to me.

By the end of day two, I had accumulated enough oomph inside my body to roll myself over, sit up on the edge of the bed with assistance, and then stand upright while leaning on someone for support. Although I found this activity to take every ounce of energy I had, I was making terrific physical progress. My right arm was very weak and continued to ache, but I could wield it around using my shoulder muscles.

On and off throughout the day, the energy in my body waxed and waned from a little bit of energy to a completely empty tank. With sleep, my reservoir filled a little and then I spent that energy trying to *do* or *think* something. Once my reserve was used up, I had to go back to sleep. I learned immediately that I had no staying power and once my energy was shot, I fell limp. I realized I had to pay very close attention to my energy gauge. I would have to learn how to conserve it and be willing to sleep to restore it.

Day two ended with a visit from Steve bearing news that

G.G. would be arriving in Boston early the next morning. Initially, I didn't understand the significance of G.G.—as I had lost the concept of what a mother was. I spent the rest of my waking moments that evening trying to piece together *Mother, Mother, Mother. G.G., G.G., G.G.* I kept repeating the words to find those files, open them, and remember. Eventually, I kind of understood what a mother was and what G.G. represented . . . enough so that I felt excited that she would be here tomorrow.

TEN

Day Three: G.G. Comes to Town

On the morning of day three, I was moved out of Neurology ICU and ended up sharing a room with a very interesting character. This woman had been suffering from epileptic seizures so the doctors had her head all wrapped up in a large white towel, with numerous electrodes and wires protruding from her head in all directions. The wires were attached to a variety of recording devices that lined her side of the room and, although she was free to move about between her bed, chair, and bathroom, she was quite the sight! I'm sure all of my visitors thought she resembled Medusa. Out of boredom, she routinely struck up a conversation with everyone who looked in on me. I, on the other hand, was desperate for silence and minimal sensory stimulation. The TV noise from her half of the room was a painful suction of my energy. I considered it totally counterintuitive to my idea of what I found to be conducive for healing.

There was a lot of excitement floating in the air that morning. My colleagues, Francine and Steve, had already arrived and several doctors were milling about the immediate area. The results were in from the angiogram and it was time to get down to the business of setting my treatment plan. I remember clearly the moment G.G. came around the corner into my room. She looked me straight in the eye and came right to my bedside. She was gracious and calm, said her hellos to those in the room, and then lifted my sheet and proceeded to crawl into bed with me. She immediately wrapped me up in her arms and I melted into the familiarity of her snuggle. It was an amazing moment in my life. Somehow she understood that I was no longer her Harvard doctor daughter, but instead I was now her infant again. She says she did what any mother would have done. But I'm not so sure. Having been born to my mother was truly my first and greatest blessing. Being born to her a second time has been my greatest fortune.

I felt perfectly content all wrapped up in my mother's love. She was kind and soft and obviously freaking out a little, but overall, I thought she was nice and I liked her. It was a perfect moment for me, and who could ask for anything more? I was catheterized so I never had to get out of bed and this very nice woman walked right into my life and surrounded me with love!

And then the conference began. Introductions were made, reports were in, and all the key players were present. Dr. Young set the tone and spoke directly to me as though I could understand. I appreciated that she did not simply speak to the others about me. First, she introduced Dr. Christopher Ogilvy, a neurosurgeon who specialized in arteriovenous malformations (AVMs). Dr. Ogilvy explained that the angiogram confirmed that my brain contained an AVM, and this congenital malformation was responsible for my hemorrhage. I had a history of migraine headaches that never responded to medication. As it turns out, my physicians predicted that I was not having migraines at all, but was experiencing small bleeds over the years.

Although I could not understand much of what was being said during this pow-wow around my bedside, I was focused on what was being conveyed nonverbally. The expressions on people's faces, the tones of their voices, how they held their bodies as they exchanged information—were fascinating to me. In a funny sort of way, I was comforted to know that the gravity of my situation did actually warrant all of this fuss. No one wants to create this much commotion only to learn that, no, it was not really a heart attack—just gas!

The atmosphere in the room was tense as Dr. Ogilvy described the problems with the blood vessels in my brain. When he suggested that I have a craniotomy to remove the remnants of the AVM and a clot the size of a golf ball, G.G. became unglued and her nervousness was obvious. Dr. Ogilvy further explained that if the AVM were not surgically removed, I stood the likely chance of hemorrhaging again and next time, I might not be so lucky getting help.

Honestly, I didn't really understand all of the details about what they were proposing to do—partly because the cells in my brain that understood language were swimming in a pool of blood and partly because of the sheer speed of their conversation. In my condition, I thought I understood that they were planning on passing a suction instrument up through my femoral artery into my brain to suck out the excess blood and threatening tangle of vessels. I was aghast when I realized it was their plan to cut my head open! Any self-respecting neuroanatomist would *never* allow anyone to cut their head open! Intuitively, if not academically, I understood that the pressure dynamics between the thoracic, abdominal, and cranial cavities are so delicately balanced that any major invasion like a craniotomy would certainly throw all of my energy dynamics completely out of whack. I feared that if they opened my head while I was already energetically compromised, I would never be able to recover my body or any of my cognition.

I made it perfectly clear to everyone that under no circumstances would I ever agree to permit them to open my head. No one seemed to understand that my body was already thoroughly deflated and I would not be able to survive another severe blow—even if it was a highly calculated one. Nevertheless, I knew I was vulnerable and at the mercy of the people in this room.

The meeting ended with the craniotomy option temporarily tabled, even though it was clear to everyone (except me) that it was now G.G.'s job to convince me to have the surgery. With tremendous compassion, G.G. intuited my fears and tried to comfort me, "That's okay sweetie, you don't have to have the surgery. No matter what, I'll take care of you. But if you don't have that AVM removed, there will always be the possibility that your brain will spit blood again. In that case, you can move in with me and I'll be attached to your hip for the rest of your life!" While my mom is a wonderful woman, living my life with her attached to my hip was not what I had in mind. Within a couple of days, I agreed to have the surgery to remove the AVM. It then became my job to get my body strong enough, over the next few weeks, to survive the pending blow.

For the next few days following the stroke, my stamina waxed and waned proportionately with my napping and exerted effort. I learned early that every effort I put forth was the only effort that was important. On day one, for example, I had to rock and rock and rock some more before I had enough oomph to roll upward. While in this stage of rocking, I had to recognize that rocking was the only activity that mattered. Focusing my success on the final goal of sitting up was not wise because it was far beyond my current ability. If I had decided that sitting up was the goal, and then tried and failed repeatedly on every trial, I would have been disappointed with my inability and stopped trying. By breaking the effort of sitting up into the smaller steps of rocking and then rolling upward, I found regular success along the way—and celebrated accordingly—with sleep. So it was my strategy to rock and then rock some more. Once I mastered rocking frequently, then I strove to rock with enthusiasm. By the time I could rock with ease, my body flowed into the next natural movement of rolling upward. And then again, my efforts were all about rolling upward, frequently, and then with enthusiastic vigor. Rolling upward with enthusiasm led me right into sitting up and I enjoyed the ongoing satisfaction of success.

Essentially, I had to completely inhabit the level of ability that I could achieve before it was time to take the next step. In order to attain a new ability, I had to be able to repeat that effort with grace and control before taking the next step. Every little *try* took time and energy, and every effort was echoed by a need for more sleep.

By day four, I was still spending most of my time sleeping as my brain craved minimal stimulation. It was not that I was depressed, but my brain was on sensory overload and could not process the barrage of incoming information. G.G. and I agreed that my brain knew best what it needed to do in order to recover. Unfortunately, it is not common for stroke survivors to be permitted to sleep as much as they would like. But for me, we felt that sleep was my brain's way of taking a "time-out" from new stimulation. We acknowledged that my brain was still physically traumatized and it was obviously totally confused concerning the information coming in through my sensory systems. We agreed that my brain needed quiet time to make sense out of what it had just experienced.

For me, sleep was filing time. You know how chaotic an office can become if you don't take time to file? It was the same
for my brain—it needed time to organize, process, and file its
hourly load.

I had to choose between physical and cognitive efforts because they both wore me out. On the physical front, I was
making terrific progress regaining my basic stability. I could
sit up with some ease now, stand and even walk a little down
the hall with lots of assistance. My voice, on the other hand,
was weak since I had no strength to expel air. As a result, I
spoke in a soft whisper and my speech was broken and labored. I struggled with finding the right word and frequently
confused meanings. I remember thinking water but saying
milk.

Cognitively, I struggled to comprehend my existence. I still
couldn't think in terms of past or future so I burned a lot of
mental energy trying to piece together my present moment.
Although thinking was very difficult for me, I was making
cognitive improvement. I had grown accustomed to my doctor telling me to remember three things, and then at the end
of our time together, asking me what those three things had
been. G.G. says that she knew I was going to be all right the
day he asked me to remember: firefighter, apple, and 33
Whippoorwill Drive. I had failed this task miserably up to
this point, but I decided that today I was going to pay attention to nothing else that he said and just repeat the words
over and over again in my mind, holding them in memory
until it was time to blurt them out. At the end of our visit, he
asked me to recall the three items. With confidence I uttered,
"Firefighter, apple, something Whippoorwill Drive." Then I
added that although I couldn't recall the exact address, I'd go
up and down the street knocking on every door until I found
the right house! G.G. breathed a huge sigh of relief when she

heard this. To her it indicated that my resourceful brain was back on track, and she was reassured that I would once again be able to find my way in the world.

On that same day, Andrew came for his daily visit and one of the games he would play with me to assess my cognitive aptitude was to ask me to count backwards from one hundred by sevens. This task was particularly difficult for me because the cells in my brain that understood mathematics had been permanently destroyed. I asked someone for the first few answers to that question and the next time Andrew asked me, I spewed three or four of the correct responses! I immediately confessed that I had cheated and really had no clue how to begin this task. But it was important to me that Andrew understand that although certain portions of my mind could not function, other parts of my brain, in this case my scheming mind, would compensate for lost abilities.

On day five, it was time for me to go home to get strong enough to endure surgery. A physical therapist taught me how to climb a stair, with support, and then I was released to G.G.'s care. I felt physically in peril as my mother drove like a Hoosier provincial in downtown Boston traffic! We draped my face to block out the sunlight. I prayed the whole way home.

ELEVEN

Healing and Preparing for Surgery

On December 15, 1996, I returned to my Winchester apartment where I now had less than two weeks to prepare for surgery. I lived on the second floor of a two-family home so I had to sit on my butt and bop myself up the stairs. (No, that's not the way the physical therapist taught me to do it!) By the time I made it up that last step, I was drained and my brain craved sleep. I was home. Finally. Home, where I could crawl into a hole and hibernate without extraneous interruptions. All of me longed for healing quietude. I collapsed on my waterbed and passed out.

I was totally blessed to have G.G. as my caregiver. If you ask her, she will tell you that she had no idea what she needed to do—she just let things unfold naturally, step-by-step. She intuitively understood that to get from A to C, I had to learn A, then B, and then C. It was as if I had an infant brain again and had to learn virtually everything from scratch. I was back to the basics. How to walk. How to talk. How to read. How to write. How to put a puzzle together. The process of physical recovery was just like stages of normal development. I had to go through each stage, master that level of ability, and then the next step unfolded naturally. Methodically, I had to learn to rock and then roll over before I could sit up. I had to sit up and rock forward before I could stand. I had to stand before I could take that first step, and I had to be relatively stable on my feet before I could climb a stair by myself.

Most important, I had to be willing to *try.* The *try* is everything. The *try* is me saying to my brain, *Hey, I value this connection and I want it to happen.* I may have to try, try, and try again with no results for a thousand times before I get even an inkling of a result, but if I don't *try,* it may never happen.

G.G. started the process of walking me by taking me back and forth between my bed and the bathroom. That was enough exercise for the day! Then it was back to sleep for another six hours! The first few days were like that. Lots of sleep, lots of energy expended to get to the bathroom or be fed, maybe a brief time for snuggling. Then it was back to sleep until the next go-around. Once I mastered the trek to the bathroom, I headed for the living room couch where I could sit up and eat some food. Learning to use a spoon with grace took some serious effort.

One of the keys to my successful recovery was that both G.G. and I were extremely patient with me. Neither of us bemoaned what I could not do; instead we always marveled at what I could do. My mother's favorite saying during moments of trauma had always been, "It could be worse!" And we both agreed, as bleak as my situation appeared on the surface, it could have been a lot worse. I have to say, G.G. was really wonderful during this process. I was the youngest of three and my mother had been a very busy woman during my toddler years. It was really sweet for me to get the chance to be mothered by her again at this level of dependence. G.G. was persistent and kind. She never raised her voice or criticized me. I was wounded and she understood that. She was warm and loving and whether I "got it" or not didn't matter. We were caught up in the process of recovery and every moment brought new hope and new possibilities.

To celebrate, Mama and I would talk about my abilities. She was superb at reminding me about what I could not accomplish yesterday and how far I had come today. She had an eagle eye for understanding what I could do and what obstacle was in my way for attaining the next level toward my goal. We celebrated all my accomplishments. She helped me clearly define what was next and helped me understand what I needed to do to get there. She kept me on track by paying attention to my details. A lot of stroke survivors complain that they are no longer recovering. I often wonder if the real problem is that no one is paying attention to the little accomplishments that are being made. If the boundary between what you can do and what you cannot do is not clearly defined, then you don't know what to try next. Recovery can be derailed by hopelessness.

I had a blow-up mattress that Mama filled with air and she built a little bedroom for herself on the floor of my living room. She took care of everything—the grocery list, the phone calls, even the bills. She was considerate and let me sleep and sleep and sleep some more. Again, we both trusted that my brain knew what it needed in order for it to mend itself. As long as I was not sleeping due to depression, we respected the healing power of sleep.

Once home, we let my brain set its own routine. I would sleep for about six hours and then be awake for about 20 minutes. Generally, the average length of time for a complete sleep cycle is 90–110 minutes. If I was awakened prematurely by external forces, I had to go back to sleep and start that cycle over. Otherwise, I would wake with a severe headache, an irritable attitude, and not be able to either sort through stimulation or focus my attention. To protect my sleep, I slept with earplugs and G.G. turned the TV and phone down low.

After a few days of intense sleeping, my energy reservoirs enabled me to remain awake for longer periods of time. Mama was a real taskmaster, and there was no wasted time or energy. When I was awake, I was a sponge for learning and she either put something in my hands for me to do or exercised my body. Yet, when I was ready to sleep, we honored that my brain had reached its maximum level of input and we put it to bed so it could rest and integrate.

Exploring life and recovering files with G.G. was fabulous fun. She learned quickly that there was no point in asking me Yes/No questions if she really wanted to know what I was thinking. It was way too easy for me to zone out about something that I didn't really care about and just B.S. her. To make sure she had my attention and I was actually working my mind, she asked me multiple-choice questions. "For lunch," she would say, "you can have minestrone soup," and then I would go on a search in my brain to figure out what minestrone soup was. Once I understood what that option was, then she would proceed with another choice. "Or, you can have a grilled cheese sandwich." Again, I would explore my brain for what a grilled cheese sandwich was. Once the image and understanding came to me, she went on. "Or, you can have tuna salad." I remember pondering *Tuna, tuna, tuna* and no image or understanding came into my mind. So I queried, "Tuna?" Mama countered, "Tuna fish from the ocean, a white meat mixed with mayonnaise, onion, and celery." Since I could not find the file for tuna salad, that's what we chose for lunch. That was our strategy if I couldn't find the old file; we made it a point to make a new one.

The telephone rang all the time and G.G. was a real trooper

at keeping everyone abreast of our daily successes. It was important that she had people to talk to about how well things were going, and it was helpful to me to have her positive attitude cheering me on. Day after day she shared stories that would remind me about how far we had come. Occasionally friends came to visit, but G.G. recognized that social exchange used up my energy reserve and left me totally drained and not interested in working. She made the executive decision that getting my mind back was more important than visitation, so she stood as the guard at my door and strictly limited my social time. TV was also a terrible energy drain, and I couldn't speak on the phone because I was completely dependent on the visual cues of lip reading. We were both respectful of what I needed to do, or not do, to recover.

Somehow we innately understood that I needed to heal my brain and challenge my neurological systems as quickly as possible. Although my neurons were stunned, technically very few of them had actually died. I would not have any official speech, occupational, or physical therapy until a couple of weeks after my surgery and in the meantime, my neurons were hungry to learn. Neurons either thrive when connected in circuit with other neurons, or they die when they sit in isolation without stimulation. G.G. and I were both highly motivated to get my brain back, so we took advantage of every moment and every precious ounce of energy.

My friend Steve had two little girls, so he brought me a collection of their books and toys. Included in the bag were children's puzzles and games. G.G. was now armed with a repertoire of age-appropriate things for me to do, and it was her policy that if I was awake and had any energy at all, she worked me.

My energy reserve did not discriminate between cognitive

versus physical activity. Energy use was energy use so we had to create a balanced strategy for recovering everything. As soon as I was able to walk around my apartment with some assistance, G.G. took me on a tour of my life. We began in the art space as I had an entire room set up for cutting stained glass. As I looked around the room, I was amazed. All of this gloriously beautiful glass! How delightful! I was an artist. And then she took me into my music room. When I strummed the strings on my guitar and then my cello, I marveled at the magic in my life. I wanted to recover.

Opening old files in my mind was a delicate process. I wondered what it would take to recall all those filing cabinets lining my brain, which contained the details of my previous life. I knew that I knew all of this stuff; I just had to figure out how to access the information again. It had been over a week since my brain had experienced the severe trauma of the hemorrhage, but the cells in my brain were still not capable of functioning correctly because of the golf ball–sized blood clot. From my perspective, I felt that every present moment was rich with experience and existed in absolute isolation. Once my back was turned, however, I was in a new rich moment and the details of the past lingered in an image or a feeling but quickly disappeared.

One morning, G.G. decided I was ready to tackle a children's puzzle so she put the puzzle box into my hands and had me look at the picture on the cover. She then helped me open the box by pulling up the lid and placed a little tray on my lap so I could dump out all the pieces. My fingers were weak and my dexterity poor, so this task would be an excellent challenge. I was very good at monkey-see, monkey-do.

G.G. explained to me that these pieces of the puzzle would

fit together to create the whole picture on the box cover. She directed me to turn all of the pieces right side up. I asked her, "What is right side up?" and she took a piece of the puzzle and showed me how to distinguish between the front and back. Once I understood the difference, I spent a little time inspecting every piece of the puzzle and eventually all twelve pieces of the puzzle were right side up. Wowie! What a sense of accomplishment! Just performing that simple mental and physical task was extremely difficult, and although I felt exhausted to have endured that level of concentration and focus, I was excited and eager to continue.

For the next task G.G. said, "Now pick out all the pieces that have an edge." I asked, "What's an edge?" Again, she patiently picked up a couple of pieces with an edge and showed me the straight cut. I then proceeded to separate out all of the edges. And once again, I felt totally accomplished and mentally fatigued.

G.G. then said to me, "I want you to take these 'outsy' pieces and hook them together with these 'insy' pieces. Also, notice that some of these insy and outsy pieces are different sizes." My right hand was extremely weak so just holding the pieces and making comparisons took a lot of effort. Mama watched me very closely and realized that I was trying to fit pieces together that obviously did not belong together based upon the image on their front side. In an effort to help me, G.G. noted, "Jill, you can use color as a clue." I thought to myself *color, color,* and like a lightbulb going off in my head, I could suddenly see color! I thought, *Oh my goodness, that would certainly make it much easier!* I was so worn out that I had to go to sleep. But the next day, I went straight back to the puzzle and put all the pieces together using color as a clue. Every day we rejoiced what I could do that I could not do the day before.

It still blows my mind (so to speak) that I could not see color until I was told that color was a tool I could use. Who would have guessed that my left hemisphere needed to be told about color in order for it to register? I found the same to be true for seeing in three dimensions. G.G. had to teach me that I could see things in different planes. She pointed out to me how some objects were closer or farther away, and that some things could be positioned in front of others. I had to be taught that items, which are positioned behind other items, may have some of their parts hidden, and that I could make assumptions about the shapes of things that I could not see in their entirety.

By the end of my first week at home, I was ambulating around my apartment pretty well and was highly motivated to find ways to exercise my body to make it stronger. One of my favorite chores, even before the stroke, was washing dishes. However, in this condition, it proved to be one of my greatest teachers. Balancing myself in front of the sink and handling delicate plates and dangerous knives was pretty challenging in and of itself, but who would guess that organizing a clean dish rack required the ability to calculate? As it turns out, the only neurons in my brain that actually died on the morning of the stroke were the ones capable of understanding mathematics. (How ironic it was that my mother had spent her entire life teaching mathematics!) I could handle washing dishes, but calculating how to get all those clean dishes to fit in that tiny little rack, well, that totally dumbfounded me! It took almost a year for me to figure it out.

I loved collecting the mail from my front box. Every day for six weeks, I received five to fifteen cards from people who were cheering me on. Although I could not read what they wrote, I would sit on G.G.'s mattress and look at the pictures, touch the cards and literally feel the love radiating from every message. G.G. would read the cards to me every afternoon. We hung them up all over the apartment so I was surrounded by all this love—on the doors, on the walls, in the bathroom, everywhere! It was really wonderful to receive these cards with the fundamental message being something like, "Dr. Jill, you don't know who I am, but I met you when you keynoted in Phoenix. Please come back to us. We love you and your work is so important to us." Every day I received this touching reinforcement of who I had been before the stroke. There is no question in my mind that it was the power of this unconditional support and love that gave me the courage to face the challenges of recovery. I will always be grateful for my friends and NAMI family who reached out to me and believed in me.

Learning to read again was by far the hardest thing I had to do. I don't know if those cells in my brain had died or what, but I had no recollection that reading was something I had ever done before, and I thought the concept was ridiculous. Reading was such an abstract idea that I couldn't believe anyone had ever thought of it, much less put forth the effort to figure out how to do it. Although G.G. was a kind taskmaster, she was insistent about my learning and placed a book titled *The Puppy Who Wanted a Boy* in my hands. Together we embarked upon the most arduous task I could imagine: teaching me to make sense of the written word. It befuddled me how she could think these squiggles were significant. I remember her showing me an "S" and saying, "This is an 'S,'" and I would say, "No Mama, that's a squiggle." And she would say, "This squiggle is an 'S' and it sounds like 'SSSSS.'" I thought the woman had lost her mind. A squiggle was just a squiggle and it made no sound.

My brain remained in pain over the task of learning to

read for quite some time. I had a real problem concentrating on something that complicated. Thinking literally was hard enough for my brain at this early stage, but jumping to something abstract was beyond me. Learning to read took a long time and a lot of coaxing. First, I had to understand that every squiggle had a name, and that every squiggle had an associated sound. Then, combinations of squiggles—er—letters, fit together to represent special combinations of sounds (sh, th, sq, etc). When we string all of those combinations of sounds together, they make a single sound (word) that has a meaning attached to it! Geez! Have you ever stopped to think about how many tiny little tasks your brain is performing this instant just so you can read this book?

Although I struggled and struggled with learning how to read again, my brain showed obvious progress each day. We celebrated when I could finally read the sounds (words) out loud, even though I displayed no comprehension. As the days went by, my recall about the overall content of the story improved and G.G. and I were both motivated to keep plodding along.

The next step, of course, was to associate a meaning to the sound. This was particularly difficult since I was already having a hard time recalling my verbal vocabulary. The blood clot was pushing against the fibers running between my two language centers, so neither of them was working properly. Broca's area in the front of my brain was having problems creating sounds while Wernicke's in the back of my brain was confusing my nouns. There seemed to be a serious gap in my information processing and often I could not articulate what I was thinking. Although I would think that I wanted a glass of water, and picture a glass of water in my mind, the word "milk" would still come out of my mouth. Although it was helpful for people to correct me, it was vitally important that no one either finish my sentences or constantly prompt me. If I were to ever regain these abilities, then I needed to find that circuitry within my mind, in my own time, and exercise it.

Day by day, I became stronger and more capable of physical exertion. The first time G.G. took me out into the yard was a fascinating learning experience. As I stood on the front walk, I needed to be taught that the lines in the cement on the sidewalk were not significant and that it was okay for me to step on those. I needed to be told that, because otherwise I didn't know. Then I needed to be taught that the line on the edge of the sidewalk was important because there was a dip there into the grass and if I was not careful, I could twist my ankle. Again, I didn't know that and I needed to be told. And then there was the grass. I needed to be shown that the texture of the grass was different from the texture of the pavement and that sinking down into the grass was okay—I just needed to pay attention and adjust my balance. G.G. let me feel what it was like to walk on snow, and she held me while my foot slipped on ice. If she was going to exercise me outside, I had to relearn that each of these textures had different features, characteristics, and their unique hazards. She kept reminding me, "What's the first thing a baby does with anything you give it?" The answer, of course, is that it puts it in its mouth to feel it. G.G. knew I needed to have direct physical contact with the world to learn kinesthetically. She was a brilliant teacher.

The upcoming surgery was going to be a huge hit to my energy, and I was committed to being physically capable of enduring it. I felt that I had lost my "brightness" when the hemorrhage occurred, and my body felt dull and weary. It seemed as if there were a veil separating me from the world outside. Dr. Young assured us that surgically removing the blood clot from my brain could potentially shift my perception and I might feel "bright" again. I figured that if I could get the brightness of my spirit back, then it didn't really matter how much I recovered, and I could be happy with whatever came my way.

My apartment was located on a busy street in Winchester, Massachusetts, and my backyard abutted a complex of apartments for the elderly. The driveway through the complex made a loop and G.G. would walk me around this natural track for exercise. I couldn't make it far in the early days, but with perseverance we eventually made it all the way around the loop. Sometimes we would loop twice if the weather permitted.

On the really cold days and the days of fresh snow, G.G. took me to the local grocery store for my daily exercise. She would go in and do her shopping and I would start walking up and down the aisles. This was a painful environment for me for several reasons. First, the intensity of the fluorescent lights was so powerful I had to constantly look down. G.G. encouraged me to wear sunglasses to block out the glare but this did little for the overpowering enormity of the room. Second, there was so much written information coming at me from all of the food items that I felt totally bombarded with stimuli. Third, the exposure to strangers was difficult for me emotionally. It was easy for others to see that I was a woman with some sort of problem. My face had that glazed-over look, and my movements were very deliberate and in slow motion when compared to the normal shopper. Many people rushed their baskets past me. Some even snarled and grumbled at me with what I interpreted as contempt. It was hard to shield myself from the negative vibrations in the environment. Occasionally, a kind spirit offered me assistance or a smile. I found facing the busy world to be intimidating and frightening.

I was introduced to the mechanics of everyday life by accompanying G.G. when she needed to do things. I became her baby duck in training, and when I had enough energy, I followed her everywhere. Who would guess that a trip to the laundromat was excellent rehabilitation? After spending time in my apartment separating the light-colored clothing from the darks, we bagged them with care. Upon arrival at the laundromat, we dumped the bags into the washers. G.G. put a quarter into my hand and then a nickel and dime. I didn't know anything about money so this was her chance to teach me. Again, the cells in my brain that understood mathematics were no longer functioning, and my attempt to deal with something so abstract as money was pitiful. When G.G. queried, "What's one plus one?" I paused for a moment, explored the contents of my mind and responded, "What's a one?" I didn't understand numbers, much less money. It felt as though I was in a foreign country with a currency I didn't understand.

Repeatedly, G.G. and I engaged in monkey-see, monkey-do behavior. The washers all ended their cycle so close to one another that I suddenly went from having nothing to do to having an overwhelming quantity to do. First we had to empty the washers. Then before loading the dryers we had to separate out the heavier items from the lighter ones. G.G. explained our strategy to me along the way. With my energy level, the washers were bearable, but frankly, the grand finale of the dryers was more demanding than I could cognitively manage! It was impossible for me to perform the "dryer dance" of pulling dry items out along the way and slamming the door closed quickly enough to keep the dryer spinning. I felt confused and desperate and wanted to crawl in a hole, hide my head, and lick my wounds. Who knew that laundry could evoke such panic in someone?

Christmas was rapidly approaching and G.G. invited my friend, Kelly, to spend the holiday with us. Together, the three of us decorated my apartment. On Christmas Eve, we found a small Christmas tree and on Christmas Day we celebrated by going out to dinner at the local Denny's. It was the simplest yet richest Christmas G.G. and I ever spent together. I was alive and recovering, and that was all that mattered.

Christmas was a day for rejoicing, but in two days I would walk into Massachusetts General Hospital to have my head cut open. From my perspective, there were two things I still needed to accomplish before surgery. One was mental and the other physical. My language was slowly coming back and it was important to me that I thank the hundreds of people who had sent me cards, letters, and flowers. I felt an intense desire to let them know I was okay, thank them for their love, and rally their continued prayers for what would come next. Folks from all over the country had signed me up for prayer lists and prayer circles ranging from local churches to the Pope's list. I felt incredible love coming my way and I wanted to share my gratitude while I still had some linguistic ability.

The greatest threat surgery posed was not only the loss of the language I had recovered, but also the loss of all future ability to ever become linguistically fluent. Since the golf ball–sized blood clot abutted the fibers running between the two language centers in my left hemisphere, it was possible that language might be excised during the surgical process. If the surgeons had to remove some of my healthy brain tissue while resecting the AVM, the consequence could be permanent loss of speech. I had come so far in my recovery that the mere possibility of this setback was chilling, but in my heart I knew that whatever the outcome, language or no language, I would still be me and we would begin again.

Although I failed miserably at reading and writing with a pen (left hemisphere/right hand), I could sit at my computer and type a simple letter (both hemispheres/both hands) that followed my stream of thought. It took me a very long time as I hunt-and-pecked at the keyboard, but somehow my body/mind connection made it happen. The most interesting thing about this experience was that after I finished typing the letter, I was not capable of reading what I had just written (left hemisphere)! G.G. edited the letter and sent it out the night following my surgery, along with a handwritten note. Since my recovery, I have heard of many stroke survivors who, although they could not speak (left hemisphere), they were capable of singing their messages (both hemispheres). I'm amazed at the resiliency and resourcefulness of this beautiful brain to find a way to communicate!

I worked day in and day out to get my body strong enough to endure the very calculated hit of surgery. Yet, there was one more task I wanted to achieve before my head met the saw. Five minutes up the street from my apartment was the Fellsway, a magnificent wooded acreage encompassing a couple of small mountainlike lakes. The Fellsway had been a magic-land for me. Most days after work, I unwound by wandering the trails among the pines, and rarely did I see another soul. I would sing and dance and prance and pray there. For me, it was a sacred place where I could commune with nature and rejuvenate.

I desperately wanted to climb that steep slippery hill up into the Fellsway before surgery. I ached to stand on top of the gigantic boulders, spread my arms in the breeze and feel the replenishment of my life force power. On the day before surgery, with Kelly by my side, I slowly climbed the hill and made my dream come true. There atop the boulders overlooking the lights of Boston, I rocked in the breeze and breathed in long, strong, empowering breaths. No matter what the next day's surgery held, this body of mine was the life force power of trillions of healthy cells. For the first time since the stroke, I felt my body was strong enough to endure the upcoming craniotomy.

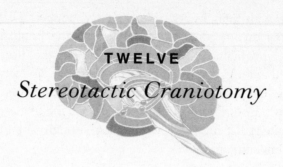

TWELVE

Stereotactic Craniotomy

At 6:00 on the morning of December 27, 1996, flanked on either side by G.G. and Kelly, I marched into Massachusetts General Hospital to have my head cut open. When I think about courage, I think about that morning.

I have had long blonde hair since I was a little girl. The last thing I remember saying to Dr. Ogilvy before he injected me with some meds was, "Hey, doc, I'm thirty-seven and single; please don't leave me totally bald!" On that note, he knocked me out.

G.G. and Kelly were quite upset with how long surgery lasted. It was late in the afternoon before word finally got to them that I was in the recovery room. Upon awakening, I realized that I felt different now. There was brightness in my spirit again and I felt happy. Up to this point, my emotions had been relatively flat. I had been observing the world, but not really engaging with it emotionally. I had missed my childlike enthusiasm since the hemorrhage and was relieved to feel like "me" again. I knew that whatever the future now held, I could face it with joy in my heart and I would be all right.

Shortly after awakening from surgery, I discovered that the left third of my head had been shaved. A nine-inch upside-down U-shaped scar—running three inches up in front of my ear, three inches horizontally over my ear, and three inches down behind my ear, was covered with an enormous patch of gauze. How nice of the good doctor to leave the right half of my head covered with hair. The moment G.G. arrived at my bedside she blurted, "Say something!" Her greatest fear, of course, was whether or not the surgeons had to take some of my language center neurons rendering me mute. I was able to speak to her softly. We both welled up in tears. The surgery had been an absolute success.

Jill's Nine-Inch Scar

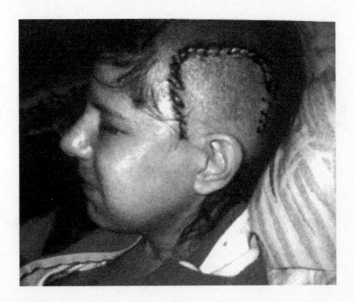

After the surgery, I stayed in the hospital for the next five days. For the first forty-eight hours, I begged to have bags of ice applied to my head. I don't know why, but my brain felt as though it were on fire and the cold ice relieved the intense heat so I could sleep.

My last night in the hospital was New Year's Eve. In the middle of the night, I sat up in the window, all alone, watching the lights of downtown Boston. I wondered what the new year would bring. I pondered the irony of my experience—a brain scientist having a stroke. I celebrated the joy I felt and the lessons I had learned. I was touched by the daunting reality: I was a stroke survivor.

THIRTEEN

What I Needed the Most

*P*lease note that I have created a list of "Recommendations
for Recovery" as a synopsis of this chapter's comments
and recommendations for how I needed to be assessed and
what I needed the most to recover. This list is located in Appendices A and B and is available for your personal use.

Recovery was a decision I had to make a million times a
day. Was I willing to put forth the effort to *try*? Was I willing
to momentarily leave my newly found ecstatic bliss to try to
understand or reengage with something in the external world?
Bottom line, was I willing to endure the agony of recovery?
At this level of information processing, I was well aware of
the difference between that which gave me pain versus that
which gave me pleasure. Being out in the la-la land of my
right hemisphere was enticing and wonderful. Trying to engage my analytical left mind was painful. Because it was a
conscious decision for me to *try*, it was critically important
that I be surrounded by competent and attentive caregivers.
Otherwise, frankly, I probably would not have bothered to
make the effort.

In order for me to choose the chaos of recovery over the

peaceful tranquility of the divine bliss that I had found in the absence of the judgment of my left mind, I had to reframe my perspective from "Why do I have to go back?" to "Why did I get to come to this place of silence?" I realized that the blessing I had received from this experience was the knowledge that deep internal peace is accessible to anyone at any time. I believe the experience of Nirvana exists in the consciousness of our right hemisphere, and that at any moment, we can choose to hook into that part of our brain. With this awareness, I became excited about what a difference my recovery could make in the lives of others—not just those who were recovering from a brain trauma, but everyone with a brain! I imagined the world filled with happy and peaceful people and I became motivated to endure the agony I would have to face in the name of recovery. My stroke of insight would be: *Peace is only a thought away, and all we have to do to access it is silence the voice of our dominating left mind.*

Recovery, however you define it, is not something you do alone, and my recovery was completely influenced by everyone around me. **I desperately needed people to treat me as though I would recover completely.** Regardless of whether it would take three months, two years, twenty years, or a lifetime, I needed people to have faith in my continued ability to learn, heal, and grow. The brain is a marvelously dynamic and ever-changing organ. My brain was thrilled with new stimulation, and when balanced with an adequate amount of sleep, it was capable of miraculous healing.

I have heard doctors say, "If you don't have your abilities back by six months after your stroke, then you won't get them back!" Believe me, this is not true. I noticed significant improvement in my brain's ability to learn and function for eight full years post-stroke, at which point I decided my mind and body were totally recovered. Scientists are well aware that the brain has tremendous ability to change its connections based upon its incoming stimulation. This "plasticity" of the brain underlies its ability to recover lost function.

I think of the brain as a playground filled with lots of little children. All of these children are eager to please you and make you happy. (What? You think I'm confusing children with puppies?) You look at the playground and note a group of kids playing kickball, another group acting like monkeys on the jungle gym, and another group hanging out by the sand box. Each of these groups of children are doing different yet similar things, very much like the different sets of cells in the brain. If you remove the jungle gym, then those kids are not going to just go away, they are going to mingle with other kids and start doing whatever else is available to be done. The same is true for neurons. If you wipe out a neuron's genetically programmed function, then those cells will either die from lack of stimulation or they will find something new to do. For example, in the case of vision, if you put a patch over one eye, blocking visual stimulation coming into the cells of the visual cortex, then those cells will reach out to the adjacent cells to see if they can contribute their efforts toward a new function. **I needed the people around me to believe in the plasticity of my brain and its ability to grow, learn, and recover.**

When it comes to the physical healing of cells, I cannot stress enough the value of getting plenty of sleep. I truly believe that the brain is the ultimate authority on what it needs to heal itself. As I mentioned earlier, for my brain, sleep was "filing time." While awake, energy stimulation poured into my sensory systems and I was rapidly burned out by photons stimulating my retinal cells and sound waves beating chaotically on my tympanic membrane. My neurons could not keep up with the demand and quickly became incapable of making sense of any incoming information. At the most elementary level of information processing, stimulation is energy, and **my brain needed to be protected, and isolated from obnoxious sensory stimulation, which it perceived as noise.**

Over the course of several years, if I didn't respect my brain's need for sleep, my sensory systems experienced agonizing pain and I became psychologically and physically depleted. I firmly believe that if I had been placed in a conventional rehabilitation center where I was forced to stay awake with a TV in my face, alert on Ritalin, and subjected to rehab on someone else's schedule, I would have chosen to zone out more and *try* less. For my recovery, it was critical that we **honor the healing power of sleep.** I know various methodologies are practiced at rehabilitation facilities around the country, yet I remain a very loud advocate for the benefits of sleep, sleep, sleep, and more sleep interspersed with periods of learning and cognitive challenge.

From the beginning, it was vitally important that my caregivers permit me the freedom to let go of my past accomplishments so I could identify new areas of interest. **I needed people to love me—not for the person I had been, but for who I might now become.** When my old familiar left hemisphere released its inhibitions over my more artistic and musically creative right hemisphere, everything shifted and I needed my family, friends, and colleagues to support my efforts at reinventing myself. At the essence of my soul, I was the same spirit they loved. But because of the trauma, my brain circuitry was different now, and with that came a shifted perception of the world. Although I looked the same and would eventually walk and talk the same as I did before the stroke, my brain wiring was different now, as were many of my interests, likes, and dislikes.

My mind was so impaired. I remember thinking, *Can they*

take away my Ph.D.? I don't remember any anatomy! I knew that I would have to find a new career that was amenable to my newly found right hemisphere gifts. Since I had always loved gardening and lawn care, I considered this a viable future option. I desperately needed people to accept me as the person I was at that moment, and permit me the freedom to evolve as a right hemisphere dominant personality. **I needed those around me to be encouraging. I needed to know that I still had value. I needed to have dreams to work toward.**

As I stated earlier, G.G. and I inherently understood that **it was essential that we challenge my brain systems immediately.** Connections in my brain had been broken and it was crucial that we restimulate them before they either died or completely forgot how to do what they were designed to do. For recovery, our success was completely dependent upon our striking a healthy balance between my awake effort and sleep downtime. For several months after surgery, I was banned from the TV, telephone, and talk radio. They did not count as legitimate relaxation time because they sapped my energy, leaving me lethargic and not interested in learning. Again, G.G. realized early to **offer me only multiple-choice questions and never ask me Yes/No questions.** Forced choice demanded that I open old files or create new ones. Yes/No questions demanded no real thought and G.G. rarely passed by a good opportunity to activate a neuron.

Because my brain had lost its ability to think linearly, I had to relearn basic personal care, including how to dress myself. I needed to be taught to put my socks on before my shoes and why. Although I couldn't remember the real function for routine household items, I was very creative in what I chose to use for what purpose. This process of exploration was exciting. Who knew a fork made a fabulous back scratcher!

My energy was limited so we had to pick and choose, very carefully every day, how I would spend my effort. **I had to define my priorities for what I wanted to get back the most and not waste energy on other things.** Although I never thought I would regain enough of my intellect to become a scientist/teacher again, I realized that I had an amazing story to tell about the beauty and resiliency of the brain—provided I could reactivate mine. I chose to focus my rehabilitation on an art project that would help recover my physical stamina, manual dexterity, and cognitive processing. For this, I decided to create an anatomically correct stained glass brain! (See book cover.) Step one required that I come up with a design. Having lost all recollection of academics, I dug out my neuroanatomy books, spread them on the floor, and pieced together an image of what I thought would make a relatively accurate (and attractive) brain. The project exercised my gross motor skills, balance and equilibrium, as well as fine motor skills for cutting and manipulating the glass. It took eight full months for me to create that first stained glass brain. When it was done, it was candy to the eye and I was motivated to make another one, which now hangs at the Harvard Brain Bank.

Several months prior to the stroke, I had booked a public presentation at Fitchburg State College. It was scheduled for April 10, which marked the four-month anniversary of my stroke. Needing a goal to work toward, I decided that this would be my first post-stroke public presentation since it was my top priority to regain fluency with language. I made the decision that I would attend the Fitchburg gig and speak for twenty minutes. It was my goal to present in such a manner that the audience would not realize that I was a stroke survivor. Although this was an ambitious undertaking, I thought it was reasonable. I embarked upon multiple strategies to accomplish this feat.

First, I had to do something about my hair! For the first few months following surgery, I was setting a new fad for hairstyle. Because the surgeons had shaved only the left third of my head, I looked quite skewed. However, if I "combed over" what remained on the right, I could hide the nine-inch scar. The fun part was trying to figure out how to disguise the new little hairs peeking out through the comb-over. It was pretty obvious that I was partially buzzed, but by April, I was sporting quite a cute little coif. I don't know if my hair gave me away that afternoon or not, nor whether anyone wondered about those two Frankenstein-like stereotactic dents in my forehead. (The stereotactic apparatus is the large halo device physicians use to hold the head perfectly still during surgery.)

I worked very hard to prepare for that Fitchburg presentation. My first challenge was to speak clearly and intelligently to an audience, and my second challenge was to be an expert about the brain. To my good fortune, I had given a major presentation that had been professionally videotaped at the NAMI national convention just a few months prior to the hemorrhage. My primary strategy for recovering my speaking skills was to watch that video over and over again. I studied how that woman (me) on the stage worked with the microphone. I watched how she held her head and body and how she walked across the stage. I listened to her voice, the melody of how she strung words together, and how altering her volume moved her audience. I learned how to do what she did by watching her. I learned how to be me again, how to act like me and walk and talk like me again, by watching that video.

As for the content and brain expert part, I learned a lot

about the brain from that presentation, but I was no expert! The videotaped presentation itself was way too much information and way over my head. I had to wonder if that's what folks in my audience thought too! I did learn, however, how to pronounce those scientific words, and after repeated viewings, I understood the story she was telling. I really enjoyed learning about brain donation, and silently wondered if G.G. would have donated my brain to science had I died on the morning of the stroke. I laughed out loud every time I heard the Brain Bank jingle and felt pangs of grief that that woman no longer existed.

In the best style I could, I put together a twenty-minute gig that I practiced day in and day out for more than a month. As long as no one interrupted me or asked any questions about the brain, I thought I could get by without anyone detecting signs of my recent stroke. Although I was rather robotic in my movements, I didn't miss a beat with my slides and I walked away from Fitchburg feeling triumphant.

Although I did not qualify for occupational or physical therapy, I spent significant time in speech therapy for four months following surgery. Speaking was less of a problem for me than reading. G.G. had already taught me the letters of the alphabet and the sounds that go with each of those squiggles, but stringing them together as words, and then adding meaning, was really more than my brain wanted to handle. Reading for comprehension was a disaster. On my first meeting with my speech therapist, Amy Rader, I was to read a story that had twenty-three facts in it. She had me read the story out loud and then answer her questions. Out of twenty-three questions I scored two right!

When I first started working with Amy, I could read the words out loud but not attach any meaning to the sounds that came out of my mouth. Eventually, I could read one word at a time, attach a meaning to that sound, and then go on to the next word. I think a lot of the problem was that I could not attach one moment to the next or think linearly. As long as every moment existed in isolation, then I could not string ideas or words together. From the inside, I felt as though the reading part of my brain was all but dead and was not interested in learning again. With the guidance of Amy and G.G., week by week, I took the steps I needed in order to achieve my goals. It was very exciting because regaining vocabulary meant regaining some of the lost files in my brain. Just trying exhausted me, but slowly, word by hard-fought word, files were opened and I was reintroduced to the life of the woman I had been. With G.G. patiently steering from the helm, I found my way back into the veiled crevices of my gray matter.

For a successful recovery, it was important that we focus on my ability, not my disability. By celebrating my achievements every day, I stayed focused on how well I was doing. I made the choice that it didn't matter if I could walk or talk or even know my name. If all I was doing was breathing, then we celebrated that I was alive—and we breathed deeper together. If I stumbled, then we could celebrate when I was upright. If I was drooling, we could celebrate swallowing! It was way too easy to focus on my disabilities because they were vast. **I needed people to celebrate the triumphs I made every day because my successes, no matter how small, inspired me.**

By the middle of January, a few weeks after surgery, my left brain language center started to come back online and talk to me again. Although I really loved the bliss of a silent mind, I was relieved to know that my left brain had the potential to recover its internal dialogue. Up to this point, I had struggled desperately to link my thoughts together and think across time. The linearity of internal dialogue helped build a foundation and structure for my thoughts.

One of the fundamental secrets to my success was that **I made the cognitive choice to stay out of my own way during the process of recovery.** An attitude of gratitude goes a long way when it comes to physical and emotional healing. I enjoyed a lot of my recovery experience as one process flowed naturally into another. I found that as my abilities increased, so did my perception of the world. Eventually I was like a toddler wanting to go out and explore—as long as my mommy wasn't too far away. I tried a lot of new things, had a lot of successes, and tried some things that I wasn't ready for yet. But I made the choice to stay out of my own way emotionally and that meant being very careful about my self-talk. It would have been really easy, a thousand times a day, to feel as though I was less than who I was before. I had, after all, lost my mind and therefore had legitimate reason to feel sorry for myself. But fortunately, my right mind's joy and celebration were so strong that they didn't want to be displaced by the feeling that went along with self-deprecation, self-pity, or depression.

Part of getting out of my own way meant that **I needed to welcome support, love, and help from others.** Recovery is a long-term process and it would be years before we would have any idea what I would get back. I needed to let my brain heal and part of that meant allowing myself to graciously receive help. Prior to the stroke, I had been extremely independent. I worked during the week as a research scientist, traveled on the weekends as the *Singin' Scientist,* and managed my home and personal affairs completely on my own. I was not comfortable accepting help, but in this state of mental incapacitation, I needed to let people do things for me. In many ways, I was fortunate that my left hemisphere was injured, for without that ego portion of my language center, I welcomed the help of others.

My successful recovery was completely dependent on my ability to break every task down into smaller and simpler steps of action. G.G. was a wizard at knowing what I needed to be able to do in order to proceed to the next level of complexity. Whether I was rocking and rolling over with enthusiasm before I could sit up, or learning that it was okay to step on cracks while walking on the sidewalk, each of these little stages determined my ultimate success.

Because I could not think linearly, **I needed everyone to assume that I knew nothing so that I could relearn everything from the beginning.** Pieces of information no longer fit together in my brain. For instance, I might not know how to use a fork and may need to be shown on several different occasions. **I needed my caregivers to teach me with patience.** Sometimes I needed them to show me something over and over again, until my body and brain could figure out what I was learning. If I didn't "get it" then it was because that part of my brain had a hole in it and could not understand or absorb the information. When people raised their voices while teaching me, I tended to shut down. Like an innocent puppy that is being yelled at, I would become afraid of that person, repelled by their energy, and tend to not trust them. It was essential that my caregivers remember that I was not deaf; my brain was simply wounded. Most important, I needed my caregivers to teach me the twentieth time with the same patience they had the first time.

I needed people to come close and not be afraid of me. I desperately needed their kindness. I needed to be touched—stroke my arm, hold my hand, or gently wipe my face if I'm drooling. Just about everyone knows someone who has had a stroke. If their language center has been disrupted, the stroke survivor can probably not carry on a conversation with visitors. I know it can be very uncomfortable for a healthy person to try to communicate with someone who has had a stroke, but **I needed my visitors to bring me their positive energy.** Since conversation is obviously out of the question, I appreciated when people came in for just a few minutes, took my hands in theirs, and shared softly and slowly how they were doing, what they were thinking, and how they believed in my ability to recover. It was very difficult for me to cope with people who came in with high anxious energy. I really needed people to take responsibility for the kind of energy they brought me. We encouraged everyone to soften their brow, open their heart, and bring me their love. Extremely nervous, anxious, or angry people were counterproductive to my healing.

One of the greatest lessons I learned was how to feel the physical component of emotion. Joy was a feeling in my body. Peace was a feeling in my body. I thought it was interesting that I could feel when a new emotion was triggered. I could feel new emotions flood through me and then release me. I had to learn new words to label these "feeling" experiences, and most remarkably, I learned that I had the power to choose whether to hook into a feeling and prolong its presence in my body, or just let it quickly flow right out of me.

I made my decisions based upon how things felt inside. There were certain emotions like anger, frustration, or fear that felt uncomfortable when they surged through my body. So I told my brain that I didn't like that feeling and didn't want to hook into those neural loops. I learned that I could use my left mind, through language, to talk directly to my brain and tell it what I wanted and what I didn't want. Upon this realization, I knew I would never return to the personality I had been before. I suddenly had much more say about how I felt and for how long, and I was adamantly opposed to reactivating old painful emotional circuits.

Paying attention to what emotions feel like in my body has completely shaped my recovery. I spent eight years watching my mind analyze everything that was going on in my brain. Each new day brought new challenges and insights. The more I recovered my old files, the more my old emotional baggage surfaced, and the more I needed to evaluate the usefulness of preserving its underlying neural circuitry.

Emotional healing was a tediously slow process but well worth the effort. As my left brain became stronger, it seemed natural for me to want to "blame" other people or external events for my feelings or circumstances. But realistically, I knew that no one had the power to make me feel anything, except for me and my brain. Nothing external to me had the power to take away my peace of heart and mind. That was completely up to me. I may not be in total control of what happens to my life, but I certainly am in charge of how I choose to perceive my experience.

FOURTEEN
Milestones for Recovery

The number one question that I am most frequently asked is, "How long did it take you to recover?" My standard response, and I don't mean to be trite, is, "Recover what?" If we define recovery as regaining access to old programs, then I am only partially recovered. I have been very fussy this time around about which emotional programs I am interested in retaining and which ones I have no interest in giving voice to again (impatience, criticism, unkindness). What a wonderful gift this stroke has been in permitting me to pick and choose who and how I want to be in the world. Before the stroke, I believed I was a product of this brain and that I had minimal say about how I felt or what I thought. Since the hemorrhage, my eyes have been opened to how much choice I actually have about what goes on between my ears.

Physical recovery from the brain surgery was minimal when compared to the task of rebuilding my mind and recovering awareness of my body. Following surgery, G.G. kept my head wound clean and the thirty-five stitches healed beautifully. The greatest challenge I faced was a problem with the left temporal mandibular joint (TMJ) of my jaw due to surgery, but with the use of a healing system called the Feldenkrais technique, it mended quickly. The scar, however, was numb for five years and I believe the three drill holes in my skull completely reknitted by year six.

My mother was a very wise caregiver, and although she was protective, she did not stand in the way of progress. In the middle of February, two months after the stroke, I took my first solo adventure into the world. My verbal language was good enough to keep me out of trouble (we hoped) and I spent minimal time out and about alone. G.G. drove me to the airport and escorted me to my seat on the plane. A friend picked me up at the other end, so I didn't have to navigate life in the big world for very long by myself. I relished this first leap out of the nest as a huge step in my pursuit of independence. Based on this success, I was encouraged to take even bigger risks.

At the three-month mark, G.G. taught me how to drive again. Operating an enormous metal box on wheels at remarkably fast speeds with a bunch of other busy people doing the same thing while they eat, drink, smoke, and, oh yes, talk on their cell phones, reminded me that I was a fragile living being and life is a precious gift. My brain was still struggling with reading, and the hardest part of learning to drive a car again was to remember to look for written signs. Obviously this was a problem. And even when I did see the sign, my comprehension was painfully slow. *So, what's that big green sign up there saying? Oh s#*?! I just passed my exit!*

By mid-March, G.G. decided I was ready to try living on my own again. Although I was far from really being back in the game, she felt that with the support of my friends, I was ready to try my wings. She reassured me that if I needed her, all I had to do was call and she would be on the first flight out. A part of me was thrilled about my increased independence. A bigger part of me was terrified.

Within a few weeks, the first major test of my readiness to resume my life was the gig in Fitchburg. That gave me something to concentrate on while I began fending for myself. My friend Julie drove me to the presentation and it went so well that I was heady with success (pun intended). Somehow I managed to not just survive, but to thrive again. I started spending time on Brain Bank responsibilities on the computer from my home. In the beginning, just a couple of hours every few days was all I could handle. Eventually I started commuting back and forth to McLean Hospital, a day or two a week. In actual fact, the commuting was more difficult than the work.

To complicate matters, following surgery my doctors insisted that I take Dilantin as a prophylactic to prevent my brain from having seizures. I had never had a seizure, but prescribing medication is a common practice when the temporal region of the brain has been surgically violated. Like a typical patient, I hated my medication because it made me feel tired and lethargic. My biggest complaint, however, was that the medicine masked my ability to know what it felt like to be me anymore. I was already a stranger to myself because of the stroke, but mix in some medication and I was even more disoriented. Because of this experience, I find that I am much more sensitive to why some people would choose insanity over the side effects of their antipsychotic medications. I was fortunate that my doctors agreed that I could take my entire dose at night before I went to bed, so by morning my mind felt much clearer. I took Dilantin for almost two full years following surgery.

At the six-month mark, I flew back home to Indiana to at-

tend my twentieth high school reunion. This was a perfect
opportunity for me to open files about my past. Two of my
best friends escorted me around, sharing stories about our
time at Terre Haute South Vigo. The timing for this reunion
was ideal. My brain had healed enough to absorb new information as well as open old files. Attending the reunion helped
me piece together memories of my youth. But again, in this
situation, because I was a stroke survivor, it was critical that I
not see myself as less than I had been before. Friends from
my past were very kind to me and I ended up having a great
time retrieving memories.

Shortly after the reunion in June, I attended the annual
NAMI convention in July. It was the end of my three-year
term on the National Board of Directors and I was officially
stepping down. I had prepared a five-minute speech to present to an audience of more than two thousand NAMI members. With guitar in hand, tears in my eyes, and gratitude in
my heart, I thanked those wonderful people for giving me
the courage to come back. I will always cherish that box of
cards they sent me, cheering me on. I know I wouldn't be
here today, in this condition, if it hadn't been for my NAMI
family.

Walking became a very important part of my routine.
When you feel like a fluid it's impossible to know where your
physical boundaries begin and end. Walking helped me become strong again, and within the course of that first year I
managed to average three miles a day several times a week. I
walked with small weights in my hands, swinging my arms
here and there, flailing them about like a wild child—but in
rhythm. I made sure I exercised all of my muscle groups—
performing shoulder girdle, shoulder, elbow, and wrist activities. Lots of people looked at me as if I was odd, but having

lost my left hemisphere ego center, I wasn't concerned with their approval or disapproval. Walking with weights helped me regain strength, balance, and posture. In addition, I worked with a friend who, through the use of massage and acupuncture, helped me identify my physical boundaries.

By the eighth month, I was back to work full time but still not completely competent either mentally or physically. There was a sluggishness about my brain that I could not shake. Unfortunately, my job description involved some complex computer database responsibilities that I knew my mind was not capable of performing. Moreover, due to the stroke, I became acutely aware of how precious little time I have here on the planet. I wanted to move back home to Indiana. Spending time with my mom and dad, while I still have them around, became a priority in my life. Fortunately, my boss agreed that I could travel for the Brain Bank as their National Spokesperson for the Mentally Ill from anywhere, and she gave me her blessing to return to Indiana.

One year after the stroke, I moved back home to the Midwest. My favorite place on earth is Bloomington, Indiana. It's the perfect-sized college town filled with interesting and creative people—and, oh yes, it's the home of Indiana University. Coming home to Indiana felt grounding to me and I knew I was exactly where I was supposed to be when my new home phone number turned out to be the exact date of my birth day, month, and year! It was one of those synchronicities in life that let me know that I was in the right place at the right time.

Post-stroke year two was spent reconstructing, as best I could recall, the morning of the stroke. I worked with a *Gestalt* therapist who helped me verbalize my right hemisphere experience of that morning. I believed that helping people understand what it felt like to experience the neurological deterioration of my mind might help caregivers better relate to stroke survivors. I also hoped that if someone read that account and then experienced any of those symptoms, they would call for help immediately. I worked with Jane Nevins and Sandra Ackerman of the The Dana Foundation on a book proposal for this story. Although our efforts were premature, I will always be grateful for their interest and assistance in helping me outline what I valued.

Eventually, when my mind was capable of learning large volumes of information again, it was time to reengage with academia. In my second post-stroke year, I was hired by Rose Hulman Institute of Technology in Terre Haute, Indiana, to teach courses in Anatomy/Physiology and Neuroscience. As I saw it, they were paying me to relearn the details of my profession. I found that although I had lost my academic terminology (left hemisphere), I still remembered what everything looked like and their relationships to one another (right hemisphere). I ended up pushing my learning ability to the limit on a daily basis and for the entire quarter, I felt like my brain would explode from overuse. I truly believe that challenging my brain in this way was exactly what it needed. Staying one lecture ahead of the students was demanding. For twelve weeks, I balanced work with appropriate sleep, and my brain performed beautifully. I will always be grateful to the Rose Hulman Department of Applied Biology and Biomechanical Engineering for their confidence in my ability to teach again.

To give you some idea of the chronology of my recovery, here is a brief summary of the highlights of my progress year by year. Prior to the stroke, I had been an avid Free Cell player (solitaire) but it was three years before I could wrap my mind around this card game again. On the physical plane, it took four years of walking with my hand weights, three miles a day, several times a week, before I could walk with a smooth rhythm. During the fourth year, my mind became capable of multitasking—even simple things like talking on the phone while boiling pasta. Up to that point, I had to do just one thing at a time, which meant everything required my full attention. And along the journey, it wasn't my style to complain. I always remembered what I had been like immediately following the stroke, and I counted my blessings and thanked my brain a thousand times a day for responding so well to my attempts to revive it. Having had a taste of the alternative, I spend a good bit of time feeling grateful for my life.

The one thing I thought I had lost forever was the ability to understand anything mathematical. To my amazement, however, by the fourth post-stroke year, my brain was ready to tackle addition again. Subtraction and multiplication came online around post-stroke year four and a half, but division eluded me until well into year five. Working with flash cards helped me drill basic math back into my brain. Now I work with the Nintendo Brain Training and Big Brain Academy programs. I think everyone over the age of forty, as well as every stroke survivor, would benefit from using this sort of brain training tool.

By the end of the fifth year, I could jump from rock to rock along the beaches in Cancún without looking at where my feet were landing. This was a significant accomplishment, because up to this point, I had to keep my eyes looking at the ground. The highlight of my sixth post-stroke year was the fulfillment of my dream of having enough oomph in my body to climb steps two at a time. Imagery has been an effective tool for regaining physical functions. I am convinced that focusing on how it feels to perform specific tasks has helped me recover them more quickly. I had dreamed of skipping up steps every day since the stroke. I held the memory of what it felt like to race up the steps with abandon. By replaying this scene over and over in my mind, I kept that circuitry alive until I could get my body and mind coordinated enough to make it reality.

Over the years, people in my professional world have been very generous and kind to me. Initially, I was afraid my colleagues might see the post-stroke me as having less value and treat me patronizingly, or perhaps even discriminate against me. Happily this has not been the case. This stroke has not only opened my eyes to the beauty and resiliency of the human brain, but also to the generosity of the human spirit. Many beautiful people have nurtured my heart, and I am grateful for all the kindnesses I have received.

Although I had been traveling part time as the *Singin' Scientist* for the Harvard Brain Bank since post-stroke year two, during my seventh year, I accepted an adjunct teaching position in the Indiana University Department of Kinesiology. In addition, teaching Gross Anatomy has always been my greatest joy, and I began volunteering in the local IU School of Medicine's Gross Anatomy lab. Revisiting the body and teaching future physicians about its miraculous design has been a thrilling privilege for me.

Also in that seventh post-stroke year, my need for sleep at night had cut back from eleven hours to nine and a half. Up until this point, in addition to a full night's sleep, I was a happy napper. For the first seven years, my dreams had been a bizarre reflection of what was going on in my brain. Instead of having dreams with people and a story line, my mind scrolled tiny unrelated bits of data. I presume this reflected how my brain pieced pixilated information together to form a complete image. It was shocking when my dreams started featuring people and a storyline again. In the beginning, the scenes were broken and nonsensical. By the end of that seventh year, however, my mind was so busy during the night that I felt little refreshment upon waking.

During the eighth year of recovery, my perception of myself finally shifted from that of being a fluid back to that of being a solid. I began slalom water skiing regularly and I believe that pushing my body as hard as I could helped solidify my brain/body connections. I confess that although I celebrate being a solid again, I really miss perceiving myself as a fluid. I miss the constant reminder that we are all *one*.

I now live what I would describe as the perfect life. I still travel for the Harvard Brain Bank as the *Singin' Scientist*. I continue to be affiliated with the Indiana University School of Medicine in Indianapolis. I regularly spend time as the consulting neuroanatomist at the Midwest Proton Radiotherapy Institute (MPRI), housed at the IU Cyclotron, where we use a precisely guided proton beam to battle cancer. To help other stroke survivors, I'm working on the creation of a virtual reality system whereby individuals can neurologically rehabilitate themselves through what I call "visually directed intention."

On the physical front, I love skiing across Lake Monroe early in the morning, and I faithfully walk around my neighborhood in the evenings. For creativity, I play in my art space creating stained glass wonders (mostly brains) and my guitar is a continual source of pleasure. I still talk to my mom every day, and as the president of the local Greater Bloomington Area NAMI affiliate, I remain active in advocacy for the mentally ill. Helping people liberate their own inner peace, joy, and magnificent beauty has become my personal agenda.

Over the years, I have had the opportunity to share my story with audiences ranging from readers of *Discover Magazine*, and Oprah Winfrey's *O Magazine*, to the *Stroke Connection Magazine* of the American Stroke Association (ASA), as well as the *Stroke Smart Magazine* of the National Stroke Association (NSA). My recovery story has been featured on PBS's *The Infinite Mind* and can still be heard on the WFIU *Profiles* program.[14] In addition, there is a wonderful PBS program titled *Understanding: The Amazing Brain*, which airs internationally. I encourage you to tune in as they did a superb job teaching about the plasticity of the brain.

14. www.indiana.edu/~wfiu/profiles.htm

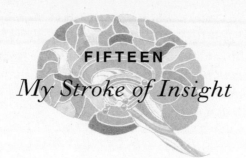

FIFTEEN

My Stroke of Insight

Having taken this unexpected journey into the depths of my brain, I am grateful and amazed that I have completely recovered physically, cognitively, emotionally, and spiritually. Over the years, the recovery of my left hemisphere skills has been tremendously challenging for many different reasons. When I lost the function of my left brain's neurological networks, I lost not only function but also a variety of personality characteristics that were apparently associated with those circuits of aptitude. Recovering cells of function that were anatomically linked to a lifetime of emotional reactivity and negative thinking has been a mind-opening experience. Although I wanted to regain my left hemisphere skills, I must say that there were personality traits that tried to rise from the ashes of my left mind that, quite frankly, were no longer acceptable to my right hemispheric sense of who I now wanted to be. From both a neuroanatomical and psychological perspective, I have had a fascinating few years.

The question I faced over and over again was, *Do I have to regain the affect, emotion, or personality trait that was neurologically linked to the memory or ability that I wanted to re-*

cover? For instance, would it be possible for me to recover my perception of my *self*, where I exist as a single, solid, separate from the whole, without recovering the cells associated with my egotism, intense desire to be argumentative, need to be right, or fear of separation and death? Could I value money without hooking into the neurological loops of lack, greed, or selfishness? Could I regain my personal power in the world, play the game of hierarchy, and yet not lose my sense of compassion or perception of equality among all people? Could I reengage with my family and not hook into my issues related to being a little sister? Most important, could I retain my newfound sense of connection with the universe in the presence of my left hemisphere's individuality?

I wondered how much of my newly found right hemisphere consciousness, set of values, and resultant personality I would have to sacrifice in order to recover the skills of my left mind. I didn't want to lose my connection to the universe. I didn't want to experience myself as a solid separate from everything. I didn't want my mind to spin so fast that I was no longer in touch with my authentic *self*. Frankly, I didn't want to give up Nirvana. What price would my right hemisphere consciousness have to pay so I could once again be judged as *normal*?

Modern neuroscientists seem satisfied intellectualizing about the functional asymmetries of our two hemispheres from a neurological perspective, but there has been minimal conversation pertaining to the psychological or personality differences contained within these two structures. Most commonly, the character of our right mind has been ridiculed and portrayed in an extremely unflattering light, simply because it does not understand verbal language or comprehend linear thought. In the case of the Dr. Jekyll and Mr. Hyde analogy, our right hemisphere personality is depicted as an uncontrollable, potentially violent, moronic, rather despicable ignoramus, which is not even conscious, and without whom we would probably be better off! In vast contrast, our left mind has routinely been touted as linguistic, sequential, methodical, rational, smart, and the seat of our consciousness.

Prior to this experience with stroke, the cells in my left hemisphere had been capable of dominating the cells in my right hemisphere. The judging and analytical character in my left mind dominated my personality. When I experienced the hemorrhage and lost my left hemisphere language center cells that defined my *self,* those cells could no longer inhibit the cells in my right mind. As a result, I have gained a clear delineation of the two very distinct characters cohabiting my cranium. The two halves of my brain don't just perceive and think in different ways at a neurological level, but they demonstrate very different values based upon the types of information they perceive, and thus exhibit very different personalities. My stroke of insight is that at the core of my right hemisphere consciousness is a character that is directly connected to my feeling of deep inner peace. It is completely committed to the expression of peace, love, joy, and compassion in the world.

This is not to say, of course, that I believe I exhibit multiple personality disorder. That is much more complicated than what I have observed. Traditionally, it has been difficult, if not impossible, for us to distinguish between our right and left mind characters simply because we experience ourselves as a single person with a single consciousness. However, with very little guidance, most people find it easy to identify these same two characters within if not themselves, then at least their parents or significant other. It is my goal to help you find a hemispheric home for each of your characters so that we can honor their identities and perhaps have more say in how we want to be in the world. By recognizing who is who inside our cranium, we can take a more *balanced-brain* approach to how we lead our lives.

It appears that many of us struggle regularly with polar opposite characters holding court inside our heads. In fact, just about everyone I speak with is keenly aware that they have conflicting parts of their personality. Many of us speak about how our head (left hemisphere) is telling us to do one thing while our heart (right hemisphere) is telling us to do the exact opposite. Some of us distinguish between what we think (left hemisphere) and what we feel (right hemisphere). Others communicate about our mind consciousness (left hemisphere) versus our body's instinctive consciousness (right hemisphere). Some of us talk about our small ego mind (left hemisphere) compared with our capital ego mind (right hemisphere), or our small self (left hemisphere) versus our inner or authentic self (right hemisphere). Some of us delineate between our work mind (left hemisphere) and our vacation mind (right hemisphere), while others refer to their researcher mind (left hemisphere) versus their diplomatic mind (right hemisphere). And of course there is our masculine mind (left hemisphere) versus our feminine mind (right hemisphere), and our yang consciousness (left hemisphere) countered by our yin consciousness (right hemisphere). And if you are a Carl Jung fan, then there's our sensing mind (left hemisphere) versus our intuitive mind (right hemisphere), and our judging mind (left hemisphere) versus our perceiving mind (right hemisphere). Whatever language you use to describe your two parts, based upon my experience, I believe they stem anatomically from the two very distinct hemispheres inside your head.

My goal during this process of recovery has been not only to find a healthy balance between the functional abilities of my two hemispheres, but also to have more say about which character dominates my perspective at any given moment. I find this to be important because the most fundamental traits of my right hemisphere personality are deep inner peace and loving compassion. I believe the more time we spend running our inner peace/compassion circuitry, then the more peace/compassion we will project into the world, and ultimately the more peace/compassion we will have on the planet. As a result, the clearer we are about which side of our brain is processing what types of information, the more choice we have in how we think, feel, and behave not just as individuals, but as collaborating members of the human family.

From a neuroanatomical perspective, I gained access to the experience of deep inner peace in the consciousness of my right mind when the language and orientation association areas in the left hemisphere of my brain became nonfunctional. The brain research performed by Drs. Andrew Newberg and the late Eugene D'Aquili[15] earlier this decade have helped me understand exactly what was going on in my brain. Using SPECT technology (single photon emission computed tomography), these scientists identified the neuroanatomy underlying our ability to have a religious or spiritual (mystical) experience. They wanted to understand which regions of the brain were involved in our capacity to undergo a shift in consciousness—away from being an individual to feeling that we are at *one* with the universe (God, Nirvana, euphoria).

Tibetan meditators and Franciscan nuns were invited to meditate or pray inside the SPECT machine. They were instructed to tug on a cotton twine when they reached either their meditative climax or felt united with God. These experiments identified shifts in neurological activity in very specific

15. *Why God Won't Go Away* (NY: Ballantine, 2001).

regions in the brain. First, there was a decrease in the activity of the left hemisphere language centers resulting in a silencing of their brain chatter. Second, there was a decrease in activity in the orientation association area, located in the posterior parietal gyrus of the left hemisphere. This region of our left brain helps us identify our personal physical boundaries. When this area is inhibited or displays decreased input from our sensory systems, we lose sight of where we begin and where we end relative to the space around us.

Orientation Association Area
(physical boundaries, space, and time)

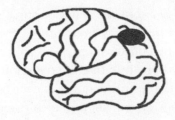

Thanks to this recent research, it makes good neurological sense that when my left language centers were silenced and my left orientation association area was interrupted from its normal sensory input, my consciousness shifted away from feeling like a solid, to a perception of myself as a fluid—at *one* with the universe.

SIXTEEN

My Right and Left Minds

I understand that no matter what information is being processed (or not being processed) in my two hemispheres, I
still experience the collective of myself as a single entity with
a single mind. I do believe that the consciousness we exhibit
is the collective consciousness of whatever cells are functioning, and that both of our hemispheres complement one another as they create a single seamless perception of the world.
If the cells and circuitry that recognize faces are functioning
correctly, then I am capable of recognizing you by your face.
If not, then I use other information to identify you, like your
voice, mannerisms, or the gait of your walk. If the cellular
circuitry that understands language is intact, then I can understand you when you speak. If the cells and circuitry that
continually remind me of who I am and where I live are destroyed, then my concept of myself will be permanently altered. That is, unless other cells in my brain learn to take over
those particular functions. Much like a computer, if I don't
have a program for word processing, then I can't perform that
function.

When we evaluate the unique characteristics of the two

cerebral hemispheres and how they process information differently, it seems obvious that they would manifest unique value systems that would consequently result in very different personalities. Some of us have nurtured both of our characters and are really good at utilizing the skills and personalities of both sides of our brain, allowing them to support, influence, and temper one another as we live our lives. Others of us, however, are quite unilateral in our thinking—either exhibiting extremely rigid thinking patterns that are analytically critical (extreme left brain), or we seldom connect to a common reality and spend most of our time "with our head in the clouds" (extreme right brain). Creating a healthy balance between our two characters enables us the ability to remain cognitively flexible enough to welcome change (right hemisphere), and yet remain concrete enough to stay a path (left hemisphere). Learning to value and utilize all of our cognitive gifts opens our lives up to the masterpiece of life we truly are. Imagine the compassionate world we could create if we set our minds to it.

Sadly, the expression of compassion is often a rarity in our society. Many of us spend an inordinate amount of time and energy degrading, insulting, and criticizing ourselves (and others) for having made a "wrong" or "bad" decision. When you berate yourself, have you ever questioned: Who inside of you is doing the yelling, and at whom are you yelling? Have you ever noticed how these negative internal thought patterns have the tendency to generate increased levels of inner hostility and/or raised levels of anxiety? And to complicate matters even more, have you noticed how negative internal dialogue can negatively influence how you treat others and, thus, what you attract?

As biological creatures, we are profoundly powerful people. Because our neural networks are made up of neurons communicating with other neurons in circuits, their behavior becomes quite predictable. The more conscious attention we pay to any particular circuit, or the more time we spend thinking specific thoughts, the more impetus those circuits or thought patterns have to run again with minimal external stimulation.

In addition, our minds are highly sophisticated "seek and ye shall find" instruments. We are designed to focus in on whatever we are looking for. If I seek red in the world then I will find it everywhere. Perhaps just a little in the beginning, but the longer I stay focused on looking for red, then before you know it, I will see red everywhere.

My two hemispheric personalities not only think about things differently, but they process emotions and carry my body in easily distinguishable ways. At this point, even my friends are capable of recognizing who's walking into the room by how I'm holding my shoulders and what's going on with that furrow in my brow. My right hemisphere is all about *right here, right now.* It bounces around with unbridled enthusiasm and does not have a care in the world. It smiles a lot and is extremely friendly. In contrast, my left hemisphere is preoccupied with details and runs my life on a tight schedule. It is my more serious side. It clenches my jaw and makes decisions based upon what it learned in the past. It defines boundaries and judges everything as right/wrong or good/bad. And oh yes, it does that thing with my brow.

My right mind is all about the richness of this present moment. It is filled with gratitude for my life and everyone and everything in it. It is content, compassionate, nurturing, and eternally optimistic. To my right mind character, there is no judgment of good/bad or right/wrong, so everything exists on a continuum of relativity. It takes things as they are and acknowledges what is in the present. The temperature is cooler today than yesterday. It doesn't care. Today it will rain. It makes no difference. It may observe that one person is taller than another person, or this person has more money than that person, but these observations are made without judgment. To my right mind, we are all equal members of the human family. My right mind does not perceive or give heed to territories or artificial boundaries like race or religion.

One of the greatest blessings I received as a result of this hemorrhage is that I had the chance to rejuvenate and strengthen my neurocircuits of innocence and inner joy. Thanks to this stroke, I have become free to explore the world again with childlike curiosity. In the absence of obvious and immediate danger, I feel safe in the world and walk the earth as though it is my backyard. In the consciousness of my right mind, we are laced together as the universal tapestry of human potential, and life is good and we are all beautiful—just the way we are.

My right mind character is adventurous, celebrative of abundance, and socially adept. It is sensitive to nonverbal communication, empathic, and accurately decodes emotion. My right mind is open to the eternal flow whereby I exist at *one* with the universe. It is the seat of my divine mind, the knower, the wise woman, and the observer. It is my intuition and higher consciousness. My right mind is ever present and gets lost in time.

One of the natural functions of my right mind is to bring me new insight in this moment so I can update old files that contain outdated information. For example, throughout my childhood I would not eat squash. Thanks to my right hemisphere, I was willing to give squash a second chance and now I love it. Many of us make judgments with our left hemisphere and then are not willing to *step to the right* (that is, into the consciousness of our right hemisphere) for a file update.

For many of us, once we have made a decision, then we are attached to that decision forever. I have found that often the last thing a really dominating left hemisphere wants is to share its limited cranial space with an open-minded right counterpart!

My right mind is open to new possibilities and thinks out of the box. It is not limited by the rules and regulations established by my left mind that created that box. Consequently, my right mind is highly creative in its willingness to try something new. It appreciates that chaos is the first step in the creative process. It is kinesthetic, agile, and loves my body's ability to move fluidly into the world. It is tuned in to the subtle messages my cells communicate via gut feelings, and it learns through touch and experience.

My right mind celebrates its freedom in the universe and is not bogged down by my past or fearful of what the future may or may not bring. It honors my life and the health of all my cells. And it doesn't just care about my body; it cares about the fitness of your body, our mental health as a society, and our relationship with Mother Earth.

The consciousness of our right mind appreciates that every cell in our bodies (except for the red blood cells) contains the exact same molecular genius as the original zygote cell that was created when our mother's egg cell combined with our father's sperm cell. My right mind understands that I am the life force power of the fifty trillion molecular geniuses crafting my form! (And it bursts into song about that on a regular basis!) It understands that we are all connected to one another in an intricate fabric of the cosmos, and it enthusiastically marches to the beat of its own drum.

Freed from all perception of boundaries, my right mind proclaims, "I am a part of it all. We are brothers and sisters on this planet. We are here to help make this world a more peaceful and kinder place." My right mind sees unity among all living entities, and I am hopeful that you are intimately aware of this character within yourself.

As much as I obviously adore the attitude, openness, and enthusiasm with which my right mind embraces life, my left mind is equally amazing. Please remember that this is the character I just spent the better part of a decade resurrecting. My left mind is responsible for taking all of that energy, all of that information about the present moment, and all of those magnificent possibilities perceived by my right mind, and shaping them into something manageable.

My left mind is the tool I use to communicate with the external world. Just as my right mind thinks in collages of images, my left mind thinks in language and speaks to me constantly. Through the use of brain chatter, it not only keeps me abreast of my life, but also manifests my identity. Via my left brain language center's ability to say, "I am," I become an independent entity separate from the eternal flow. As such, I become a single, a solid, separate from the whole.

Our left brain truly is one of the finest tools in the universe when it comes to organizing information. My left hemisphere personality takes pride in its ability to categorize, organize, describe, judge, and critically analyze absolutely everything. It thrives in its constant contemplation and calculation. Regardless of whether or not my mouth is running, my left mind stays busy theorizing, rationalizing, and memorizing. It is a perfectionist and an amazing housekeeper of corporation or home. It constantly says, "Everything has a place and everything belongs in its place." Our right mind character values humanity, while our left mind character concerns itself with finances and economy.

On the scale of *doing*, my left mind is a magnificent

multitasker and loves performing as many functions as it can at the same time. It is a true busy bee and partially measures its value by how many things it crosses off my daily to-do list. Because it thinks sequentially, it is great at mechanical manipulation. Its ability to focus on differences and distinguishing characteristics makes it a natural builder.

My left brain is particularly gifted at identifying patterns. As a result, it is adept at processing large volumes of information quickly. To keep up with life's experiences in the external world, my left mind processes information remarkably fast—much faster than my right hemisphere, which in comparison tends to hoe-de-doe along. At times my left mind may become manic, while my right mind has the potential to become lazy.

This difference in speed of thought, information processing, and output as thought, word, or deed, between our two hemispheres, is in part linked to their unique abilities to process different types of sensory information. Our right brain perceives the longer wavelengths of light. As a result, the visual perception of our right mind is somewhat blended or softened. This lack of edge perception enables it to focus on the bigger picture of how things relate to one another. Similarly, our right mind tunes in to the lower frequencies of sound that are readily generated by our body gurgles and other natural tones. Consequently, our right mind is biologically designed to readily tune in to our physiology.

In contrast, our left brain perceives the shorter wavelengths of light, increasing its ability to clearly delineate sharp boundaries. As a result, our left mind is biologically adept at identifying separation lines between adjacent entities. At the same time, our left hemisphere language centers tune in to the higher frequencies of sound, which help them detect, discriminate, and interpret tones commonly associated with verbal language.

One of the most prominent characteristics of our left brain is its ability to weave stories. This storyteller portion of our left mind's language center is specifically designed to make sense of the world outside of us, based upon minimal amounts of information. It functions by taking whatever details it has to work with, and then weaves them together in the form of a story. Most impressively, our left brain is brilliant in its ability to make stuff up, and fill in the blanks when there are gaps in its factual data. In addition, during its process of generating a story line, our left mind is quite the genius in its ability to manufacture alternative scenarios. And if it's a subject you really feel passionate about, either good or awful, it's particularly effective at hooking into those circuits of emotion and exhausting all the "what if" possibilities.

As my left brain language centers recovered and became functional again, I spent a lot of time observing how my storyteller would draw conclusions based upon minimal information. For the longest time I found these antics of my storyteller to be rather comical. At least until I realized that my left mind full-heartedly expected the rest of my brain to believe the stories it was making up! Throughout this resurrection of my left mind's character and skills, it has been extremely important that I retain the understanding that my left brain is doing the best job it can with the information it has to work with. I need to remember, however, that there are enormous gaps between what I know and what I think I know. I learned that I need to be very wary of my storyteller's potential for stirring up drama and trauma.

In the same vein, as my left brain enthusiastically manufactured stories that it promoted as the truth, it had a tendency to be redundant—manifesting loops of thought patterns that reverberated through my mind, over and over again. For many of us, these loops of thought run rampant and we find ourselves habitually imagining devastating possibilities. Unfortunately, as a society, we do not teach our children that they need to *tend carefully the garden of their minds.* Without structure, censorship, or discipline, our thoughts run rampant on automatic. Because we have not learned how to more carefully manage what goes on inside our brains, we remain vulnerable to not only what other people think about us, but also to advertising and/or political manipulation.

The portion of my left mind that I chose not to recover was the part of my left hemisphere character that had the potential to be mean, worry incessantly, or be verbally abusive to either myself or others. Frankly, I just didn't like the way these attitudes felt physiologically inside my body. My chest felt tight, I could feel my blood pressure rise, and the tension in my brow would give me a headache. In addition, I wanted to leave behind any of my old emotional circuits that automatically stimulated the instant replay of painful memories. I have found life to be too short to be preoccupied with pain from the past.

During the process of recovery, I found that the portion of my character that was stubborn, arrogant, sarcastic, and/or jealous resided within the ego center of that wounded left brain. This portion of my ego mind held the capacity for me to be a sore loser, hold a grudge, tell lies, and even seek revenge. Reawakening these personality traits was very disturbing to the newly found innocence of my right mind. With lots of effort, I have consciously chosen to recover my left mind's ego center without giving renewed life to some of those old circuits.

SEVENTEEN
Own Your Power

I define responsibility (response-ability) as the ability to choose how we respond to stimulation coming in through our sensory systems at any moment in time. Although there are certain limbic system (emotional) programs that can be triggered automatically, it takes less than ninety seconds for one of these programs to be triggered, surge through our body, and then be completely flushed out of our bloodstream. My anger response, for example, is a programmed response that can be set off automatically. Once triggered, the chemical released by my brain surges through my body and I have a physiological experience. Within ninety seconds from the initial trigger, the chemical component of my anger has completely dissipated from my blood and my automatic response is over. If, however, I remain angry after those ninety seconds have passed, then it is because I have *chosen* to let that circuit continue to run. Moment by moment, I make the choice to either hook into my neurocircuitry or move back into the present moment, allowing that reaction to melt away as fleeting physiology.

The really exciting news about acknowledging my right

and left characters is that I always have an alternative way of looking at any situation—is my glass half full or half empty? If you approach me with anger and frustration, then I make the choice to either reflect your anger and engage in argument (left brain), or be empathic and approach you with a compassionate heart (right brain). What most of us don't realize is that we are unconsciously making choices about how we respond all the time. It is so easy to get caught up in the wiring of our pre-programmed reactivity (limbic system) that we live our lives cruising along on automatic pilot. I have learned that the more attention my higher cortical cells pay to what's going on inside my limbic system, the more say I have about what I am thinking and feeling. By paying attention to the choices my automatic circuitry is making, I own my power and make more choices consciously. In the long run, I take responsibility for what I attract into my life.

Nowadays, I spend a whole lot of time thinking about thinking just because I find my brain so fascinating. As Socrates said, "The unexamined life is not worth living." There has been nothing more empowering than the realization that I don't have to think thoughts that bring me pain. Of course there is nothing wrong with thinking about things that bring me pain as long as I am aware that I am choosing to engage in that emotional circuitry. At the same time, it is freeing to know that I have the conscious power to stop thinking those thoughts when I am satiated. It is liberating to know that I have the ability to choose a peaceful and loving mind (my right mind), whatever my physical or mental circumstances, by deciding to *step to the right* and bring my thoughts back to the present moment.

More often than not, I choose to observe my environment through the nonjudgmental eyes of my right mind, allowing me to retain my inner joy and remain detached from emotionally charged circuitry. I alone decide if something will have a positive or negative influence on my psyche. Recently, I was driving down the road singing at the top of my lungs with my favorite Ginger Curry[16] CD, "I got JOOOOOY in my heart!" To my chagrin, I was pulled over for speeding (apparently there was way too much enthusiasm going on behind the wheel!). Since receiving that ticket, I have had to make the decision at least a hundred times to not feel down about it. This little voice of negativity kept trying to raise its ugly head and bum me out. It wanted to mull the drama over and over again in my mind, from every angle, but regardless of my contemplation, the situation would have the same outcome. Frankly, I find this sort of mental obsessing by my left hemisphere storyteller to be a waste of time and emotionally draining. Thanks to my stroke, I have learned that I can own my power and stop thinking about events that have occurred in the past by consciously realigning myself with the present.

With that said, however, there are some occasions when I will choose to step into the world as a single, solid, ego center separate from you. Sometimes it is just pure satisfaction for me to bump my left hemisphere stuff and attitudes up against your left hemisphere stuff and attitudes, in argument or passionate debate. More often than not, I don't like how aggression feels inside my body so I shy away from hostile confrontation and choose compassion.

For me, it's really easy to be kind to others when I remember that none of us came into this world with a manual about how to get it all right. We are ultimately a product of our biology and environment. Consequently, I choose to be compassionate with others when I consider how much painful emotional baggage we are biologically programmed to carry

16. www.gingercurry.com

around. I recognize that mistakes will be made, but this does not mean that I need to either victimize myself or take your actions and mistakes personally. Your stuff is your stuff, and my stuff is my stuff. Feeling deep inner peace and sharing kindness is always a choice for either of us. Forgiving others and forgiving myself is always a choice. Seeing this moment as a perfect moment is always a choice.

EIGHTEEN

Cells and
Multidimensional Circuitry

My good friend Dr. Jerry Jesseph lives his life by the philosophy "Peacefulness should be the place we begin rather than the place we try to achieve." I interpret this to mean that we should stem from the peaceful consciousness of our right mind and use the skills of our left mind to interact with the external world. He has also coined the phrase "dual interpenetrating awarenesses" to describe the relationship between the two halves of our brain. I think this is a profound and accurate perspective. Thanks to our corpus callosum, our two hemispheres are so intricately interwoven that we perceive ourselves as a single individual. However, through our understanding that we have two very distinct ways of being in the world, we can deliberately choose to have much more power over what's going on inside our brains than we ever imagined!

My left brain became competent again when it regained the ability to process information at fast rates of speed. Now that it is completely back online, it tends to reengage with life at what feels like a million miles an hour. Needless to say, the natural competition between my left hemisphere language centers and my right hemisphere's experience of inner peace has grounded me back in the normal human condition. A part of me is thrilled to be so functional again. A bigger part of me is terrified.

This experience of losing my left brain has opened my mind to look more positively at people who have experienced various forms of brain trauma. I often wonder, in the absence of language or the ability to communicate with others in a normal way, what insights or abilities has that person gained? I don't feel sorry for people who are different from me or perceived as not normal anymore. I realize that pity is not an appropriate response. Instead of feeling repelled by someone who is different, I am drawn toward them with kindness and curiosity. I am fascinated by their uniqueness and compelled to establish a meaningful connection, even if it is merely direct eye contact, a kind smile, or appropriate touch.

When I take responsibility for the circumstances of my life, I put myself in the driver's seat and own my power. In an attempt to maintain my sanity (peaceful heart) in a world that often feels like it spins dangerously fast, I continue to work very hard to maintain a healthy relationship between what is going on in my right and left minds. I love knowing that I am simultaneously (depending on which hemisphere you ask) as big as the universe and yet merely a heap of star dust.

Everyone's brain is different but let me share with you some of the simple things I have found to be true for mine. It seems that the more aware I am about how I am influencing the energies around me, the more say I have in what comes my way. To monitor how things are going in my life, I pay very close attention to how things are flowing, or not flowing in the world around me. Depending on what I am attracting,

I take responsibility for how things are going and consciously make adjustments along the way. This does not mean that I am in complete control of everything that happens to me. However, I am in control of how I choose to think and feel about those things. Even negative events can be perceived as valuable life lessons, if I am willing to *step to the right* and experience the situation with compassion.

Now that my left mind's language centers and storyteller are back to functioning normally, I find my mind not only spins a wild tale but has a tendency to hook into negative patterns of thought. I have found that the first step to getting out of these reverberating loops of negative thought or emotion is to recognize when I am hooked into those loops. For some of us, paying attention to what our brain is saying to us comes naturally. Many of my college students, however, complain vehemently that it takes way too much mental effort for them to simply observe what their brain is telling them. Learning to listen to your brain from the position of a nonjudgmental witness may take some practice and patience, but once you master this awareness, you become free to step beyond the worrisome drama and trauma of your storyteller.

When I become conscious of what cognitive loops my brain is running, I then focus on how these loops feel physiologically inside my body. Do I feel alert? Are my eyes dilated? Is my breath deep or shallow? Do I feel tightness in my chest? Do I feel lightness in my head? Is my stomach upset? Do I feel antsy or anxious? Are my legs jiggling? Neuronal loops (circuits) of fear, anxiety, or anger, can be triggered by all sorts of different stimulation. But once triggered, these different emotions produce a predictable physiological response that you can train yourself to consciously observe.

When my brain runs loops that feel harshly judgmental,

counterproductive, or out of control, I wait ninety seconds for the emotional/physiological response to dissipate and then I speak to my brain as though it is a group of children. I say with sincerity, "I appreciate your ability to think thoughts and feel emotions, but I am really not interested in thinking these thoughts or feeling these emotions anymore. Please stop bringing this stuff up." Essentially, I am consciously asking my brain to stop hooking into specific thought patterns. Different people do it differently of course. Some folks just use the phrase, "Cancel! Cancel!" or they exclaim to their brain, "Busy! I'm too busy!" Or they say, "Enough, enough, enough already! Knock it off!"

Simply thinking these thoughts with my inner authentic voice, however, is often not enough for me to get the message across to my storyteller, who is invested in performing its normal function. I have found that when I attach an appropriate feeling to these phrases, and think them with genuine affect, my storyteller is more amenable to this type of communication. If I'm really having trouble getting my brain to listen, I add a kinesthetic component to my message like waggling my pointed finger in the air, or standing firm with my hands on my hips. A scolding mother is more effective when she says what she means with passion and communicates her message multidimensionally.

I wholeheartedly believe that 99.999 percent of the cells in my brain and body want me to be happy, healthy, and successful. A tiny portion of the storyteller, however, does not seem to be unconditionally attached to my joy, and is excellent at exploring thought patterns that have the potential to really derail my feeling of inner peace. This group of cells has been called many things; some of my favorites include the Peanut Gallery, the Board of Directors, and the Itty Bitty S#*?!y Committee. These are the cells in our verbal mind that are totally resourceful in their ability to run our loops of doom and gloom. These cells tap into our negative attributes of jealousy, fear, and rage. They thrive when they are whining, complaining, and sharing with everyone about how awful everything is.

In extreme situations of cellular disregard, I use my authentic voice to put my language center's Peanut Gallery on a strict time schedule. I give my storyteller full permission to whine rampantly between 9:00–9:30 A.M. and then again between 9:00–9:30 P.M. If it accidentally misses *whine time*, it is not allowed to reengage in that behavior until its next allotted appointment. My cells quickly get the message that I am serious about not hooking into those negative loops of thought—but only if I am persistent and determined enough to pay attention to what circuits are running in my brain.

I'm a devout believer that paying attention to our self-talk is vitally important for our mental health. In my opinion, making the decision that internal verbal abuse is not acceptable behavior is the first step toward finding deep inner peace. It has been extremely empowering for me to realize that the negative storyteller portion of my brain is only about the size of a peanut! Just imagine how sweet life was when those cranky cells were silent. Recovering my left mind has meant that I have had to give voice to all of my cells again. However, I have learned that in order to protect my overall mental health, it is necessary for me to tend the garden of my mind and keep these cells in check. I have found that my storyteller simply needs a little disciplining directive from my conscious mind about what I want versus what I find unacceptable. Thanks to our open line of communication, my authentic self has much more say over what is going on with this particular group of cells; and I spend very little time hooked into unwanted or inappropriate thought patterns.

Having said that, however, I am often humored by the scheming antics of my storyteller in response to this type of directive. I have found that just like little children, these cells may challenge the authority of my authentic voice and test my conviction. Once asked to be silent, they tend to pause for a moment and then immediately reengage those forbidden loops. If I am not persistent with my desire to think about other things, and consciously initiate new circuits of thought, then those uninvited loops can generate new strength and begin monopolizing my mind again. To counter their activities, I keep a handy list of three things available for me to turn my consciousness toward when I am in a state of need: 1) I remember something I find fascinating that I would like to ponder more deeply, 2) I think about something that brings me terrific joy, or 3) I think about something I would like to do. When I am desperate to change my mind, I use such tools.

I have also found that when I am least expecting it—feeling either physically tired or emotionally vulnerable—those negative circuits have a tendency to raise their hurtful heads. The more aware I remain about what my brain is saying and how those thoughts feel inside my body, the more I own my power in choosing what I want to spend my time thinking about and how I want to feel. If I want to retain my inner peace, I must be willing to consistently and persistently *tend the garden of my mind* moment by moment, and be willing to make the decision a thousand times a day.

Our patterns of thought are grounded in rich multidimensional circuitry that we can learn to scrutinize. First, each thought pattern has a subject—something that I am cognitively thinking about. For example, let's say I am thinking about my little dog, Nia, who spent a good share of her last eight years sitting on my lap, helping me write this book. Thinking about Nia is a specific circuit in my brain. Second, each thought pattern may or may not be accompanied by an adjoining emotional circuit of which I am cognizant. In the case of Nia, I generally experience great joy when I think of her because she was a marvelously loving creature. In my brain, the subject circuitry of Nia and the emotional circuitry of joy are intimately linked. Finally, these specific circuits of thought and emotion may also be linked with some of my more complex physiological circuitry, which upon stimulation would result in predictable behavior.

For example: when I think about Nia (thought circuitry) I experience the feeling of joy (emotional circuitry) and more often than not, I experience great excitement (physiological circuitry) and engage in puppylike behavior (multidimensional circuitry). I instantly revert to a childlike voice and my eyes dilate. My joy becomes palpable, and I spontaneously waggle my body as if I were wagging my tail! Yet, in addition to this circuitry of excitement and animation, on other occasions I am also inclined to respond to the thought of Nia with consuming sadness—as I mourn the loss of my beloved four-legged friend. In the instant of a shifted thought, and its underlying emotional and physiological circuitry, my eyes might well up with tears. Caught in the loop of deep grief, my chest tightens, my breathing becomes shallow, and emotionally I feel depressed. Feeling weak in the knees, my energy wanes and I succumb to loops of darkness.

These passionate thoughts and feelings have the potential to jump instantly into my mind, but again, after their ninety seconds have come and gone, I have the power to consciously choose which emotional and physiological loops I want to hook into. I believe it is vital to our health that we pay very close attention to how much time we spend hooked into the circuitry of anger, or the depths of despair. Getting caught up in these emotionally charged loops for long periods of time can have devastating consequences on our physical and mental well-being because of the power they have over our emotional and physiological circuitry. However, with that said, it is equally important that we honor these emotions when they surge through us. When I am moved by my automatic circuitry, I thank my cells for their capacity to experience that emotion, and then I make the choice to return my thoughts to the present moment.

Finding the balance between *observing* our circuitry and *engaging* with our circuitry is essential for our healing. Although I celebrate my brain's ability to experience all of my emotions, I am cautious about how long I remain hooked into running any particular loop. The healthiest way I know how to move through an emotion effectively is to surrender completely to that emotion when its loop of physiology comes over me. I simply resign to the loop and let it run its course for ninety seconds. Just like children, emotions heal when they are heard and validated. Over time, the intensity and frequency of these circuits usually abate.

Really powerful thoughts are perceived as powerful because they simultaneously run multiple circuits of emotion and physiology. Thoughts that we would define as neutral are perceived as neutral because they are not stimulating complex circuitry. Paying attention to which array of circuits we are concurrently running provides us with tremendous insight into how our minds are fundamentally wired, and consequentially, how we can more effectively tend our garden.

In addition to spending a lot of time conversing with my brain cells, I'm having a big love-fest with the fifty trillion molecular geniuses making up my body. I am so grateful that they are alive and working together in perfect harmony that I implicitly trust them to bring me health. The first thing every morning and the last thing every night, I faithfully hug my pillow, wrap one hand in the other, and consciously thank my cells for another great day. I care enough to say it out loud. "Thank you girls. Thanks for another great day!" and I say it with an intense feeling of gratitude in my heart. I then implore my cells, *Please, heal me*, and I visualize my immune cells responding.

I unconditionally love my cells with an open heart and grateful mind. Spontaneously throughout the day, I acknowledge their existence and enthusiastically cheer them on. I am a wonderful living being capable of beaming my energy into the world, only because of them. When my bowels move, I cheer my cells for clearing that waste out of my body. When my urine flows, I admire the volume my bladder cells are capable of storing. When I'm having hunger pangs and can't get to food, I remind my cells that I have fuel (fat) stored on my hips. When I feel threatened, I thank my cells for their ability to fight, flee, or play dead.

At the same time, I listen to my body when it speaks to me. If I feel tired, I give my cells sleep. When I'm feeling slothlike, I give my cells movement. When I'm in pain, I become quiet, coddle the wound, and consciously surrender into the pain, which helps it dissipate. Pain is the tool our cells use to communicate to our brain that there is trauma somewhere in our body. Our cells stimulate our pain receptors in order to get our brain to focus and pay attention. Once my brain acknowledges the existence of the pain, then it has served its purpose and either lightens up in intensity, or goes away.

From my perspective, the focused human mind is the most powerful instrument in the universe, and through the use of language, our left brain is capable of directing (or impeding) our physical healing and recovery. My verbal left ego mind functions as the head cheerleader of my fifty trillion molecular geniuses, and when I periodically encourage my cells with *You go girls!*, I can't help but think this induces some sort of vibration within my body that promotes a healing environment. I'm a believer that when my cells are healthy and happy, I am healthy and happy.

None of this is to say that people with true mental illness have the capacity to choose completely what is going on inside their brains. However, I do believe that all symptoms of severe mental illness stem from a biological basis: which cells are communicating with which cells, with which chemicals and in what quantities of those chemicals. Brain research is on the cutting edge of understanding the neurocircuitry underlying mental illness, and as our knowledge grows, there will be a greater understanding about how we can help people more effectively monitor and tend to the health of their minds.

For treatment options, we have the capacity to influence our brain cells chemically through the use of medication, electrically via electrical stimulation, and cognitively through psychotherapy. In my opinion, the purpose of medical treatment is to increase our ability to share a common reality. I am in favor of people exploring what resources might help them more readily connect with others. Unfortunately, 60 percent of people diagnosed with schizophrenia do not recognize that they are ill. As a result, they do not seek or value treatment and often engage in self-medication through the abuse of drugs or alcohol. Even the recreational use of these substances (by anyone) decreases our ability to share a common reality, and can thus be counterproductive to one's health.

Although some individuals advocate for the right to be insane, I am of the opinion that it is everyone's civil right to experience sanity and share in a common reality—whatever the cause of their brain illness or trauma.

NINETEEN

Finding Your Deep Inner Peace

This stroke of insight has given me the priceless gift of knowing that deep inner peace is just a thought/feeling away. To experience peace does not mean that your life is always blissful. It means that you are capable of tapping into a blissful state of mind amidst the normal chaos of a hectic life. I realize that for many of us, the distance between our thinking mind and our compassionate heart sometimes feels miles apart. Some of us traverse this distance on command. Others of us are so committed to our hopelessness, anger, and misery that the mere concept of a peaceful heart feels foreign and unsafe.

Based upon my experience with losing my left mind, I wholeheartedly believe that the feeling of deep inner peace is neurological circuitry located in our right brain. This circuitry is constantly running and always available for us to hook into. The feeling of peace is something that happens in the present moment. It's not something that we bring with us from the past or project into the future. Step one to experiencing inner peace is the willingness to be present in the right here, right now.

The more aware we are of when we run our loop of deep inner peace, the easier it is for us to purposely choose to hook into that circuitry. Some of us struggle in our attempts to recognize when we are running this circuitry only because our minds are distracted by other thoughts. This makes sense since our Western society honors and rewards the skills of our "doing" left brain much more than our "being" right brain. Thus, if you are having difficulty accessing the consciousness of your right mind circuitry, then it is probably because you have done a stupendous job learning exactly what you were taught while growing up. Congratulate your cells for their successes, and realize that, as my good friend Dr. Kat Domingo proclaims, "Enlightenment is not a process of learning, it is a process of unlearning."

Since both of our hemispheres work together to generate our perception of reality on a moment-by-moment basis, we are exercising our right mind all the time. Once you learn to recognize the subtle feelings (and physiology) running through your body when you are connected to the circuitry of the present moment, you can then train yourself to reactivate that circuitry on demand. I'm going to share with you a variety of ways I *re-mind* myself back into the consciousness and personality of my right here, right now, peaceful right brain.

The first thing I do to experience my inner peace is to remember that I am part of a greater structure—an eternal flow of energy and molecules from which I cannot be separated (see Chapter Two). Knowing that I am a part of the cosmic flow makes me feel innately safe and experience my life as heaven on earth. How can I feel vulnerable when I cannot be separated from the greater whole? My left mind thinks of me as a fragile individual capable of losing my life. My right mind realizes that the essence of my being has eternal life.

Although I may lose these cells and my ability to perceive this three-dimensional world, my energy will merely absorb back into the tranquil sea of euphoria. Knowing this leaves me grateful for the time I have here as well as enthusiastically committed to the well-being of the cells that constitute my life.

In order to come back to the present moment, we must consciously slow down our minds. To do this, first decide you are not in a hurry. Your left mind may be rushing, thinking, deliberating, and analyzing, but your right mind is very m-e-l-l-o-w.

Right now, besides reading this book, what are you doing? Are you running any cognitive loops in addition to your reading? Are you watching the clock or sitting in a busy place? Become aware of your extraneous thoughts, thank them for their service, and ask them to be silent for a little while. We're not asking them to go away, just to push the *pause button* for a few minutes. Rest assured, they're not going anywhere. When you are ready to reengage with your storyteller again, it will jump right back online.

When we are hooked into cognitive thoughts and running mental loops, technically we are not in the present moment. We can be thinking about something that has already occurred or about something that has not yet happened, and although our body is right here, right now, our mind is somewhere else. In order to come back to the experience of the present moment, allow your consciousness to shift away from those cognitive loops that distract you from what is happening right now.

If you will, think about your breathing. Since you are reading this book, then you are probably sitting in a relaxed state. Draw in a big deep breath. Go ahead, it's okay. Pull air deep into your chest and watch your belly swell. What's going on inside your body? Is it in a comfortable position? Is your stomach feeling calm or queasy? Are you hungry? How full is your bladder? Do you have a dry mouth? Do your cells feel tired or refreshed? How is your neck? Just take a pause from any distracting thoughts and observe your life for a moment. Where are you sitting? How's the lighting? How do you feel about where you are sitting? Take another deep breath, and now another. Relax into your body—soften your jaw and that furrow in your brow. Revel in the fact that in this moment, you are a living, thriving human being! Let that feeling of celebration and gratitude flood your consciousness.

To help me find my way back into my peaceful right mind, I look at how my body organizes information into systems and capitalize on those already established circuits. I find that paying attention to sensory information as it streams into my body is a very helpful tool. However, I don't just focus on the sensory information, I consciously hook into the physiological experience underlying that sensory circuitry. I ask myself repeatedly, *How does it feel to be here doing this?*

Eating, drinking, and being merry is something that happens in the present moment. Our mouths contain various types of sensory receptors permitting us the ability to not only taste different flavors but to perceive unique textures and varied temperatures. Try observing more closely how different foods taste. Pay attention to the textures of different foods and how they feel in your mouth. What foods would you classify as fun food and why? I love chasing around those little individual gelatin balls in tapioca pudding. Spaghetti is a great texture to play with too. The most fun I have with food, however, is squishing the guts out of half-frozen peas, or smooshing mashed potatoes between my teeth! I realize your mother probably expunged these behaviors out of your dining repertoire when you were young, but in the privacy of your own home, I'm thinking it's probably okay. It's really hard to entertain stress-inducing thoughts when you're having fun with food!

Besides the physical attributes of food consumption, it is vitally important that we consider the physiological impact food has on our body and mind. Beyond the traditional focus on nutritional value, try paying attention to how select foods make your body feel. Both sugar and caffeine have me crawling out of my skin within minutes after I consume them. It's a feeling I don't like and thus try to avoid. Eating foods that contain the chemical tryptophan (milk, bananas, and turkey) rapidly increase the levels of the neurotransmitter serotonin in my brain and cause me to feel mellow. I purposely choose these foods when I want to concentrate and feel calm.

In general, carbohydrates turn immediately into sugar and make my body feel lethargic and my brain spastic. Also, I don't like the way carbs spike my sugar/insulin response and then leave me craving. I like the way proteins charge me up and give me energy without stimulating emotional highs and lows. You may have a different response to these foods, and that's okay. A balanced diet is important, but paying attention to how you burn energy and how foods make you feel inside your skin should be a top priority.

One of the easiest ways to shift just about anyone's mood (for better or for worse) is through stimulation of their nose. If you are overly sensitive, life in the real world can be unbearable. Capitalizing on our noses to shift ourselves back into the present moment is easy. Light a scented candle and let vanilla, rose, or almond lift you up beyond your recollections of stress. When random smells waft past you, hook into that cognitive loop and spend quality time trying to identify the scent. Score it on a scale of one to ten for pleasure or yuck. Remember to feel the physiology that underlies differ-ent scents. Let them move you into the here and now.

If you are having a problem with your ability to smell, then I'm a true believer that unless the circuits have been per-manently severed, it is possible to increase your sensitivity. When you purposely pay attention to the smells around you, you're sending a message to your brain saying you value that connection. If you want to improve your sense of smell, spend more time sniffing different scents and talk to your cells! Let them know you want them to improve their ability. If you are willing to change your behavior such that you spend more time consciously thinking about what you are smelling, and you're willing to focus your mind on the act of smelling, then the neuronal connections will get reinforced and potentially become stronger.

When it comes to vision, there are basically two ways you can use your eyes. Take a moment right now to look at the view in front of you. What do you see? Your right mind takes in the big picture. It sees the view as a whole where every-thing is relative. It observes the entire expanse and does not focus on any of the details. Your left mind immediately fo-cuses in on the contour of individual objects and delineates the specific entities making up the view.

When I stand on a mountaintop and let my eyes relax, my right mind takes in the magnificence of the open vista. Physi-ologically, I feel the majesty of the overall view deep inside my being, and I am humbled by how beautiful our planet is. I can recall this moment by either reconstructing the vision or by recalling the feeling it elicits. My left mind is completely different. It eagerly focuses my attention on the specific types of trees, the colors in the sky, and analyzes the sounds of specific birds. It discriminates the types of clouds, delineates the tree line, and registers the temperature of the air.

Right now, take a pause from your reading. Close your eyes and identify three sounds you hear. Go ahead. Relax your mind and expand your perception. What do you hear? Listen close and listen far. As I sit here in Dipper Cabin at the Rocky Ridge Music Center in the Rocky Mountains near Estes Park, my ears are privy to the gurgling sounds of a creek as it passes right outside my picture window. When I focus my mind on distant sounds, I hear bits and pieces of classical music as children practice their instruments. Focusing my ears up close, I hear the hum of the heater, right here in the cabin, as it warms me.

Listening to music that you love, in the absence of cognitive analysis or judgment, is another great way to come back to the here and now. Let sound move you not just emotionally but physically. Allow your body to rock and sway or dance and play in accordance with the rhythm. Surrender your inhibitions and let your body get caught in the flow.

Of course, the absence of sound can be equally as beautiful. I love putting my ears under water in the tub to create a space of sound deprivation. I also focus on my body's gurgles when they occur and send my cells praise for their ongoing efforts. I have found that my mind is easily distracted by too much auditory stimulation, so I often work, or travel, with earplugs. I believe that preventing stimulation overload in my brain is my responsibility, and earplugs have been a true sanity saver on many occasions.

Our largest and most diverse sensory organ is our skin. Just as our brain runs various circuits that think, experience emotion, or involve specific combinations of physiological re-activity, our skin is stippled with very specific receptors capable of detecting very specific forms of stimulation. As with our other senses, we are all unique in how sensitive we are to light touch, pressure, heat and cold, vibration, and pain. Some of us adapt more quickly than others. Although most of us don't spend much time thinking about our clothes after we put them on, some of us remain so sensitive that our minds obsess over their texture or weight. I thank my cells regularly for their ability to adapt to incoming stimulation. Imagine how preoccupied our minds would be all the time if we couldn't.

Humor me again, if you will, and take another pause from your reading. This time, close your eyes and think about the information you are currently detecting from your skin. How is the temperature of the air? What is the texture of your clothing—soft or scratchy, light or heavy? Is anything pushing up against you—maybe a pet or a pillow? Just think about your skin for a moment. Can you feel your watch, or those glasses on your nose? How about your hair draping on your shoulder?

From a therapeutic perspective, there is perhaps nothing more intimate than touch, be it physical connection with another human, a furry friend, or even your household plants. The physical benefits of nurturing and being nurtured are priceless. Simply taking a shower and feeling the water splashing upon your body is a great way to jolt yourself back into the present moment. Feeling the pressure of water against your skin, by taking a bath or playing in a pool, is excellent light pressure and temperature stimulation. Allow these forms of activity the power to lull you back into the here and now. Train yourself to pay closer attention to when your different circuits are stimulated. As you do, you encourage them to function.

Deep body massage is also great for a number of reasons. Not only does it help relieve tension in your muscles, but it also increases the movement of the fluids in your cellular environment. The internal world of your body is how your cells obtain nutrition and clear their waste. I enthusiastically support any type of stimulation that increases their standard of living.

One of my favorite ways of using touch to come back to the here and now is through raindrops. Walking in the rain is a multidimensional experience that moves me deeply. Drops of water spattering on my face instantly shift me into the beauty and innocence of my right mind as I feel enveloped by a deep sense of purification. Feeling the warmth of sunshine upon my face or the kiss of a breeze on my cheek also connects me directly with a part of myself that feels at *one* with all that is. I absolutely love standing on the ocean's edge with my arms spread wide, flying in the breeze. By remembering the smells, sounds, tastes, and how I felt deep inside, I can transport myself back to Nirvana in an instant.

The more attention we pay to the details of how things look, sound, taste, smell, feel against our skin and feel physiologically inside our body, the easier it is for our brain to recreate any moment. Replacing unwanted thought patterns with vivid imagery can help us shift our consciousness back toward our deep inner peace. Although it is great to use our senses to rebuild an experience, I believe the real power in experiential recreation is located in our ability to remember what the underlying physiology feels like.

It's impossible for me to end this section on the use of sensory stimulation to bring one into the present moment, and not touch on the subjects of energy dynamics and intuition. For those of you who have very sensitive right hemispheres, I know you understand what I am talking about. At the same time, I appreciate that for many of us, if our left mind cannot smell it, taste it, hear it, see it, or touch it, then we are skeptical as to whether or not it exists. Our right brain is capable of detecting energy beyond the limitations of our left mind because of the way it is designed. I hope your level of discomfort about such things as energy dynamics and intuition has decreased as you have increased your understanding about the fundamental differences in the way our two hemispheres collaborate to create our single perception of reality.

Remembering that we are energy beings designed to perceive and translate energy into neural code may help you become more aware of your own energy dynamics and intuition. Can you sense the mood of a room when you first walk in? Ever wonder why you seem to be content one minute and then fraught with fear the next? Our right hemisphere is designed to perceive and decipher the subtle energy dynamics we perceive intuitively.

Since the stroke, I steer my life almost entirely by paying attention to how people, places, and things feel to me energetically. In order to hear the intuitive wisdom of my right mind, however, I must consciously slow my left mind down so I am not simply carried along on the current of my chatty storyteller. Intuitively, I don't question why I am subconsciously attracted to some people and situations, and yet repelled by others. I simply listen to my body and implicitly trust my instincts.

At the same time, my right mind completely honors the phenomenon of cause and effect. In a world of energy, where everything influences everything, it seems naïve for me to disregard the insights of my right mind. If I am shooting a bow and arrow, for example, I don't just focus on the target's bull's-eye, but I trace the path between the arrow tip and the center of the target. I visualize the perfect amount of force exerted by my muscles as they pull back the arrow, and focus my mind on the fluidity of the process rather than the finality of the end product. I find that when my perception is expanded and I imagine the experience, my accuracy is increased. If you are involved in sports, you have the power to choose how you want to perceive yourself in relationship to your target or goal. You can see yourself as separate—you positioned at spot A and your target at spot Z, or you can see yourself at *one* with the target and in the flow with all the atoms and molecules in the space between.

Our right brain perceives the big picture and recognizes that everything around us, about us, among us, and within us is made up of energy particles that are woven together into a universal tapestry. Since everything is connected, there is an intimate relationship between the atomic space around and within me, and the atomic space around and within you—regardless of where we are. On an energetic level, if I think about you, send good vibrations your way, hold you in the light, or pray for you, then I am consciously sending my energy to you with a healing intention. If I meditate over you or lay my hands upon your wound, then I am purposely directing the energy of my being to help you heal. How the arts of Reiki, Feng Shui, acupuncture, and prayer (to mention only a few) work remain pretty much medical mysteries. This is mostly because our left brains and science have not yet successfully caught up with what we understand to be true about how our right hemisphere functions. However, I believe our right minds are perfectly clear about how they intuitively perceive and interpret energy dynamics.

Shifting away from the subject of sensory systems, we can also use the skills of our motor output systems to shift our perspective into the here and now. Purposely relaxing muscles you routinely hold tense can help you release pent-up energy and feel better. I'm constantly checking in with the tension in my forehead and inevitably, if I can't fall asleep at night, I consciously loosen my jaw and then proceed to pass right out. Thinking about what is going on with your muscles is a great way to pull your mind back into the present. Systematically squeezing and relaxing them may help you come back to the here and now.

Lots of people use movement and exercise to shift their minds. Yoga, Feldenkrais, and Tai Chi are awesome tools for personal development, relaxation, and growth. Noncompetitive sports are also a great way to get you back into your body and out of your left brain. Walking in nature, singing, creating, and playing music, or getting lost in the arts can easily shift your perspective back to the present moment.

Another avenue for shifting one's focus away from the churning loops of our left cognitive mind is through purposely using our voice to interrupt those looping patterns of thought that we find distressful or distracting. I find that using repetitious sound patterns such as mantra (which literally means "place to rest the mind") is very helpful. By breathing deeply and repeating the phrase *In this moment I reclaim my JOY,* or *In this moment I am perfect, whole, and beautiful,* or *I am an innocent and peaceful child of the universe,* I shift back into the consciousness of my right mind.

Listening to a verbal meditation that guides me into a thought pattern with emotion and physiology is another great way to shift my mind away from unwanted loops. Prayer, whereby we use our mind to intentionally replace unwanted thought patterns with a chosen set of thought patterns, is another way to consciously guide one's mind away from the incessant squirrel cage of verbal repetition into a more peaceful place.

I absolutely love vocal tuning with sounding bowls. These are large bowls made of exquisite quartz crystal. When stroked, the bowls resonate so powerfully that I can feel the vibration right down to my bones. My worries don't stand a chance at hanging on to my mind when the sounding bowls are in play.

I also draw Angel Cards[17] several times a day to help me stay focused on what I believe is important in life. The original Angel Cards come in sets of assorted sizes with each card having a single word written on them. Every morning when I first get up, I ritualistically invite an angel into my life and draw a card. I then focus my attention on that particular angel throughout my day. If I am feeling stressed or have an important phone call to make, I will often draw another angel to help me shift my mind. I am always in quest of being open to what the universe will bring me. I use the Angel Cards to shift me back into a state of being generous of spirit, as I really like what I attract when I am open. Some of the angels include: enthusiasm, abundance, education, clarity, integrity, play, freedom, responsibility, harmony, grace, and birth. Drawing angels is one of the simplest and most effective tools I have found to help me shift my mind out of my left hemisphere's judgment.

If I had to pick one output (action) word for my right mind, I would have to choose *compassion*. I encourage you to ask yourself, what does it mean to you to be compassionate? Under what circumstances are you inclined to be compassionate and what does compassion feel like inside your body?

Generally, most of us are compassionate with those we see as our equals. The less attached we are to our ego's inclination for superiority, the more generous of spirit we can be with others. When we are being compassionate, we consider another's circumstance with love rather than judgment. We

17. www.innerlinks.com

see a homeless person or a psychotic person and approach them with an open heart, rather than fear, disgust, or aggression. Think about the last time you reached out to someone or something with genuine compassion. How did it feel inside your body? To be compassionate is to move into the right here, right now with an open heart consciousness and a willingness to be supportive.

If I had to choose one word to describe the feeling I feel at the core of my right mind, I would have to say *joy*. My right mind is thrilled to be alive! I experience a feeling of awe when I consider that I am simultaneously capable of being at *one* with the universe, while having an individual identity whereby I move into the world and manifest positive change.

If you have lost your ability to experience joy, rest assured the circuitry is still there. It is simply being inhibited by more anxious and/or fearful circuitry. How I wish you could lose your emotional baggage, just like I did, and shift back into your natural state of joy! The secret to hooking into any of these peaceful states is the willingness to stop the cognitive loops of thought, worry, and any ideas that distract us from the kinesthetic and sensory experience of being in the here and now. Most important, however, our desire for peace must be stronger than our attachment to our misery, our ego, or our need to be right. I love that old saying, "Do you want to be right, or do you want to be happy?"

Personally, I really like the way happy feels inside my body and therefore I choose to hook into that circuitry on a regular basis. I've often wondered, *If it's a choice, then why would anyone choose anything other than happiness?* I can only speculate, but my guess is that many of us simply do not realize that we have a choice and therefore don't exercise our ability to choose. Before my stroke, I thought I was a product of my brain and had no idea that I had some say about how I

responded to the emotions surging through me. On an intellectual level, I realized that I could monitor and shift my cognitive thoughts, but it never dawned on me that I had some
say in how I perceived my emotions. No one told me that it
only took ninety seconds for my biochemistry to capture, and
then release me. What an enormous difference this awareness
has made in how I live my life.

Another reason many of us may not choose happiness is
because when we feel intense negative emotions like anger,
jealousy, or frustration, we are actively running complex circuitry in our brain that feels so familiar that we feel strong
and powerful. I have known people who consciously choose
to exercise their anger circuitry on a regular basis simply because it helps them remember what it feels like to be themselves.

It is just as easy for me to habitually run the happiness circuit as it is for me to run the anger circuit. In fact, from a biological perspective, happiness is the natural state of being for
my right mind. As such, this circuitry is constantly running
and is always available for me to tap into. My anger circuit, on
the other hand, does not always run, but can be triggered
when I experience some sort of threat. As soon as the physiological response has passed out of my bloodstream, I can
resume my joy.

Ultimately, everything we experience is a product of our
cells and their circuitry. Once you have tuned in to how different circuits feel inside your body, then you can pick and
choose how you want to be in the world. I, personally, feel
allergic to how fear and/or anxiety feel in my body. When
these emotions surge through me, I feel so uncomfortable
that I want to crawl out of my skin. Because I don't like the

way these emotions feel physiologically, I'm not inclined to hook into that circuitry on a regular basis.

My favorite definition of fear is "False Expectations Appearing Real," and when I allow myself to remember that all of my thoughts are merely fleeting physiology, I feel less moved when my storyteller goes haywire and my circuitry is triggered. At the same time, when I remember that I am at *one* with the universe, then the concept of fear loses its power. To help protect myself from having a trigger-happy anger or fear response, I take responsibility for what circuitry I purposely exercise and stimulate. In an attempt to diminish the power of my fear/anger response, I intentionally choose not to watch scary movies or hang out with people whose anger circuitry is easily set off. I consciously make choices that directly impact my circuitry. Since I like being joyful, I hang out with people who value my joy.

As I mentioned earlier, physical pain is a physiological phenomenon that is specifically designed to alert our brain that tissue damage has occurred somewhere in our body. It's important we realize that we are capable of feeling physical pain without hooking into the emotional loop of suffering. I am reminded of how courageous little children can be when they become extremely ill. Their parents may hook into the emotional circuitry of suffering and fear, while the child seems to be adapting to his illness without the same negative emotional drama. To experience pain may not be a choice, but to suffer is a cognitive decision. When children are ill, it is often more difficult for the child to handle parental grief than it is for the child to endure the illness.

The same can be true for anyone who is ill. Please be very careful what circuits you stimulate when you visit someone who is not well. Death is a natural process we all must experience. Just realize that deep inside your right mind (deep within your heart's consciousness) rests eternal peace. The easiest way I have found to humble myself back into a state of peaceful grace is through the act of gratitude. When I am simply grateful, life is simply great!

TWENTY

Tending the Garden

I have learned so much from this experience with stroke that I actually feel fortunate to have taken this journey. Thanks to this trauma, I have had the chance to witness first-hand a few things about my brain that otherwise I would never have imagined to be true. For these simple insights, I will always be grateful—not just for myself but for the hope these possibilities may bring to how we, as a people, choose to view and nurture our brains and consequently behave on this planet.

I am grateful for your willingness to join me on this intense journey. I sincerely hope that whatever circumstances brought you to this book, you move forward having gleaned some insight into your brain or the brain of another. I trust with my right hemisphere's heart consciousness that this book will now flow from your hands into the hands of someone who may benefit from it.

I always end my e-mails with a tagline quote from Einstein. I believe he got it right when he said, "I must be willing to give up what I am in order to become what I will be." I learned the hard way that my ability to be in the world is completely dependent on the integrity of my neurocircuitry. Cell by beautiful cell, circuit by neurocircuit, the consciousness I experience within my brain is the collective awareness established by those marvelous little entities as they weave together the web I call my mind. Thanks to their neural plasticity, their ability to shift and change their connections with other cells, you and I walk the earth with the ability to be flexible in our thinking, adaptable to our environment, and capable of choosing who and how we want to be in the world. Fortunately, how we choose to be today is not predetermined by how we were yesterday.

I view the garden in my mind as a sacred patch of cosmic real estate that the universe has entrusted me to tend over the years of my lifetime. As an independent agent, I and I alone, in conjunction with the molecular genius of my DNA and the environmental factors I am exposed to, will decorate this space within my cranium. In the early years, I may have minimal input into what circuits grow inside my brain because I am the product of the dirt and seeds I have inherited. But to our good fortune, the genius of our DNA is not a dictator, and thanks to our neurons' plasticity, the power of thought, and the wonders of modern medicine, very few outcomes are absolute.

Regardless of the garden I have inherited, once I consciously take over the responsibility of tending my mind, I choose to nurture those circuits that I want to grow, and consciously prune back those circuits I prefer to live without. Although it is easier for me to nip a weed when it is just a sprouting bud, with determination and perseverance, even the gnarliest of vines, when deprived of fuel, will eventually lose its strength and fall to the side.

The mental health of our society is established by the mental health of the brains making up our society, and I must admit that Western civilization is a pretty challenging environment for my loving and peaceful right hemisphere character to live in. Obviously, I'm not alone in feeling this way, as I look at the millions of beautiful people in our society who have chosen to escape our common reality by self-medicating themselves with illicit drugs and alcohol.

I think Gandhi was right when he said, "We must be the change we want to see in the world." I find that my right hemisphere consciousness is eager for us to take that next giant leap for mankind and *step to the right* so we can evolve this planet into the peaceful and loving place we yearn for it to be.

Your body is the life force power of some fifty trillion molecular geniuses. You and you alone choose moment by moment who and how you want to be in the world. I encourage you to pay attention to what is going on in your brain. Own your power and show up for your life. Beam bright!

And when your life force wanes, I hope you will give the gift of hope and donate your beautiful brain to Harvard.

RESOURCES

American Stroke Foundation
Stroke Activity Center
5960 Dearborn
Mission, KS 66202
Phone: 913-649-1776
Toll Free: 1-866-549-1776
Fax: 913-649-6661
www.americanstroke.org

American Stroke Association
National Center
7272 Greenville Avenue
Dallas, TX 75231
Phone: 1-888-4-STROKE
www.strokeassociation.org

National Stroke Association
9707 E. Easter Lane
Centennial, CO 80112
Phone: 1-800-787-6537
www.stroke.org

To learn more, please visit www.drjilltaylor.com
and www.mystrokeofinsight.com.

RECOMMENDATIONS FOR RECOVERY

Appendix A

Ten Assessment Questions

1. Have you had my eyes and ears checked to make sure you know what I can see and hear?
2. Can I discriminate color?
3. Do I perceive three dimensions?
4. Do I have any sense of time?
5. Can I identify all of my body parts as mine?
6. Can I discriminate a voice from background noise?
7. Can I access my food? Can my hands open the containers? Do I have the strength and dexterity to feed myself?
8. Am I comfortable? Am I warm enough? Or thirsty? Or in pain?
9. Am I oversensitive to sensory stimulation (light or sound)? If so, bring me earplugs so I can sleep, and sunglasses so I can keep my eyes open.
10. Can I think linearly? Do I know what socks and shoes are? Do I know that my socks go on before my shoes?

Appendix B

Forty Things I Needed Most

1. I am not stupid, I am wounded. Please respect me.

2. Come close, speak slowly, and enunciate clearly.

3. Repeat yourself—assume I know nothing and start from the beginning, over and over.

4. Be as patient with me the twentieth time you teach me something as you were the first.

5. Approach me with an open heart and slow your energy down. Take your time.

6. Be aware of what your body language and facial expressions are communicating to me.

7. Make eye contact with me. I am in here—come find me. Encourage me.

8. Please don't raise your voice—I'm not deaf, I'm wounded.

9. Touch me appropriately and connect with me.

10. Honor the healing power of sleep.

11. Protect my energy. No talk radio, TV, or nervous visitors! Keep visitation brief (five minutes).

12. Stimulate my brain when I have any energy to learn something new, but know that a small amount may wear me out quickly.

13. Use age-appropriate (toddler) educational toys and books to teach me.

14. Introduce me to the world kinesthetically. Let me feel everything. (I am an infant again.)

15. Teach me with monkey-see, monkey-do behavior.

16. Trust that I am trying—just not with your skill level or on your schedule.

17. Ask me multiple-choice questions. Avoid Yes/No questions.

18. Ask me questions with specific answers. Allow me time to hunt for an answer.

19. Do not assess my cognitive ability by how fast I can think.

20. Handle me gently, as you would handle a newborn.

21. Speak to me directly, not about me to others.

22. Cheer me on. Expect me to recover completely, even if it takes twenty years!

23. Trust that my brain can always continue to learn.

24. Break all actions down into smaller steps of action.

25. Look for what obstacles prevent me from succeeding on a task.

26. Clarify for me what the next level or step is so I know what I am working toward.

27. Remember that I have to be proficient at one level of function before I can move on to the next level.

28. Celebrate all of my little successes. They inspire me.

29. Please don't finish my sentences for me or fill in words I can't find. I need to work my brain.

30. If I can't find an old file, make it a point to create a new one.

31. I may want you to think I understand more than I really do.

32. Focus on what I can do rather than bemoan what I cannot do.

33. Introduce me to my old life. Don't assume that because I cannot play like I used to play that I won't continue to enjoy music or an instrument, etc.

34. Remember that in the absence of some functions, I have gained other abilities.

35. Keep me familiar with my family, friends, and loving support. Build a collage wall of cards and photos that I can see. Label them so I can review them.

36. Call in the troops! Create a healing team for me. Send word out to everyone so they can send me love. Keep them abreast of my condition and ask them to do specific things to support me—like visualize me being able to swallow with ease or rocking my body up into a sitting position.

37. Love me for who I am today. Don't hold me to being the person I was before. I have a different brain now.

38. Be protective of me but do not stand in the way of my progress.

39. Show me old video footage of me doing things to re-
    mind me about how I spoke, walked, and gestured.

40. Remember that my medications probably make me
    feel tired, as well as mask my ability to know what it
    feels like to be me.

So . . . You Always Wanted to Go to HARVARD!

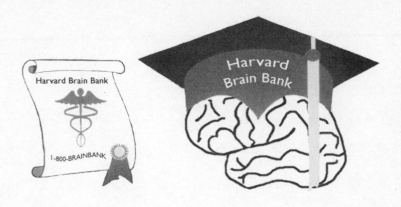

1-800-BrainBank!

Oh, I am a brain banker,
Banking brains is what I do.
I am a brain banker,
Asking for a deposit from you!

Don't worry, I'm in no hurry!
Have you considered the contribution you can make
When you are heaven bound, your brain can hang around,
To help humanity, find the key to
Unlock this thing we call insanity.
Just dial 1-800-BrainBank for information please,
Educate then donate, it's free!

Oh, I am a brain banker,
Banking brains is what I do.
I am a brain banker,
Asking for a deposit from you!

The Harvard Brain Bank is federally funded to acquire and distribute specific types of tissue for qualified investigators to do their research. Right now there is a long-term shortage of tissue donated by individuals with a psychiatric illness, their immediate family members (siblings, parents, children), as well as individuals who would be diagnosed as normal controls. These are individuals who do not have either a neurological or a psychiatric diagnosis, or have an immediate family member with a psychiatric diagnosis.

Unfortunately, stroke survivors (including myself) and anyone else with a history of brain cancer do NOT qualify for brain donation to the Harvard Brain Bank. If you are interested in some type of donation, please consider a full body donation to your local or state medical university.

If you are interested in participating in studies while you are still alive, please investigate possibilities at the National Institute of Neurological Disease and Stroke (NINDS).

INDEX